Acceptable Genes?

SUNY series on Religion and the Environment

Harold Coward, editor

Acceptable Genes?

Religious Traditions and Genetically Modified Foods

Edited by
CONRAD G. BRUNK
and
HAROLD COWARD

Published by State University of New York Press, Albany

© 2009 State University of New York

For information, contact State University of New York Press, Albany, NY
www.sunypress.edu

Production by Eileen Meehan
Marketing by Anne M. Valentine

Library of Congress Cataloging-in-Publication Data

Acceptable genes? : religious traditions and genetically modified foods / edited by
    Conrad G. Brunk and Harold Coward.
       p. cm. — (SUNY series on religion and the environment)
    Includes bibliographical references and index.
    ISBN 978-1-4384-2895-6 (hardcover : alk. paper)
    ISBN 978-1-4384-2894-9 (pbk. : alk. paper)
    1. Genetically modified foods—Religious aspects.   I. Brunk, Conrad G.
(Conrad Grebel), 1945–   II. Coward, Harold G.

    TP248.65.F66A23 2009
    215'.7—dc22   2009007283

Contents

Introduction

Conrad G. Brunk and Harold Coward

A lively debate about genetically modified foods has engaged around the world since their first introduction onto the markets of many countries in the last decades of the twentieth century. The debate has been especially intense in Europe, Japan, and parts of Africa and has led in many instances to moratoria on the introduction of genetically modified crops into the agriculture of the societies and strict requirements for the labeling of genetically modified foods and food ingredients produced in or imported into the country.

This debate has been uncharacteristically subdued in North America, where these products were first grown for commercial use and sent to markets for consumption. Public concern or opposition was limited primarily to small, often marginalized, environmental or consumer groups but did not become widespread as in other regions. One reason for this may have been that government regulators in Canada and the United States approved these products for the market with no public announcement that they were doing so and certainly without any prior public consultation, in contrast to the practice in most European countries. Indeed, most people in North America have been until very recently completely unaware that much of the food they are purchasing is from genetically modified corn, canola, soybeans, and other crops, and that genetically modified or cloned food animals have been developed and applications for their market approval submitted to their regulators.

Although public awareness is now more widespread in North America, levels of concern over GM food are still fairly low on the public's list of political priorities. One concern, however, is not low—that of the desire for labeling of these products in order to give consumers a choice whether or not to purchase them.

There are many surveys of what, for example, Canadian, American, or British populations think about various aspects of biotechnology and genetically modified foods. But these global surveys are rarely conducted in ways that capture the specific concerns of subpopulations whose attitudes or values might vary significantly from those of the population in general, such as those who are adherents of specific religious or ethical cultures and traditions. Indeed, it is well known that the manner in which people respond to requests for their opinions on nearly any subject is influenced significantly by many variables. Most important among these is the manner in which the question posed by the questioner is contextualized. For example, most subjects occupy different social roles, and it may not be evident to them which one they are being asked to adopt when responding to a question—that of consumer, citizen, parent, or member of a cultural or religious community. These identifications can make a significant difference in the way people respond to requests for their opinion, because the particular personal or social identification makes more salient to them the values and concerns that are appropriate to that identification.

Clearly, among the most important social and personal identities in terms of which people reflect their most profound values are their religious and ethical identities. Asking someone in his or her capacity as a religious believer or practitioner to advance an opinion on a matter from the point of view of that religious belief or practice is likely to elicit a different response than if one had asked for the opinion in abstraction from that role or from the point of view of some other role, such as the role of citizen or consumer. The very general question, "Do you think genetically modified foods are a good thing?" or even the more specific question, "Do you think genetically modified foods should be labeled?" asked out of any context and abstracted from any background of information that could be crucial to the subject's assessment of the issue may not elicit a reliable expression of the subject's considered judgment in light of his or her core values. A question that asks the subject to reflect on a matter from the point of view of those core values and in light of otherwise unavailable information that might be relevant to those values would be more likely to elicit a reliable response. So, the question, "As a Muslim who follows the discipline of Islamic dietary rules (*halal*), how do you feel about eating food that contains DNA from an animal that is not acceptable for you to eat?" draws the subject's attention to aspects of the issue that may not have occurred to him or her at all if the more general question had been asked or if the question had not been addressed to the subject *as a Muslim*.

The aim of this book is to understand the moral and religious attitudes of significant subpopulations within pluralistic societies whose traditions and beliefs raise for them unique questions about food and dietary practice that potentially influence their attitudes toward various types of food biotechnology. We have focused our study on those subpopulations who identify with long-standing religious and ethical traditions with well-articulated philosophies or theologies about what is appropriate to eat or how the food one eats should be produced or prepared for human consumption. Such philosophies are well represented in most of the great world religions and in the long-standing secular moral tradition of ethical vegetarianism. For the purposes of this study we have focused on those traditions with the most significant representation in North American society: Christianity, Judaism, Islam, Hinduism, Buddhism, Chinese religion, and ethical vegetarianism. All these traditions have within them specific prohibitions on the use of certain animals or plants in the diet or norms governing the cultivation or preparation of foods.[1] So all of them would seem to have the potential for raising questions about the propriety of growing or consuming certain kinds of genetically modified plants or animals, including those plants or animals that might contain DNA artificially transferred from other plants or animals considered morally problematic within the tradition. In this book we limit our focus to food consumption issues and do not consider the environmental or ecological concerns that GM foods may raise. Such concerns are left for others to address.

There is another reason why it is important to look carefully at the attitudes of well-defined religious and ethical communities toward a controversial technology such as GM foods. It is precisely because these attitudes are the product of fundamental *religiously* or *conscientiously* held moral beliefs that they have a social and legal standing that may raise them to a higher level of significance for industry and regulators than other consumer preferences. Fundamental religious and moral values do not affect consumer and citizen behavior in the same way as mere preferences, and they also carry with them moral and legal claims for respect and tolerance that mere preferences do not, especially in societies committed to the legal protection of religious liberty and freedom of conscience. This issue is discussed in more detail in the final chapter of this book.

There are at least two very different ways one might go about trying to shed useful light on a complex issue such as how different religious traditions might view challenges posed by food biotechnology. One way would be to ask expert interpreters of the religion to examine the nature and philosophical/theological basis of the religion's norms around production, preparation, and consumption of food and to explain the reasonable or expected implications of these beliefs and norms for genetically modified foods. This is a largely *normative* approach to the issue in the sense that it asks an expert who understands well the logic and rationale of the belief system behind a religiously motivated dietary practice to interpret how that belief system *ought* to view aspects of the new technology. Its strength is that it provides at least one account of how the tradition is most likely to handle the issue (if it is consistent). It also provides significant guidance to members of the tradition who sincerely wish to know how they should view the matter consistently with their fundamental beliefs or practices.

This approach also has obvious and significant weaknesses and limitations, especially for predicting the actual attitudes and behavior of those who espouse the religious or ethical outlook. One is asking for the opinion of *one expert* in the tradition. With respect to most religious or ethical philosophies, expert interpreters, whether adherents or observers, are likely to reflect widely divergent and incompatible viewpoints on the issue. Of some traditions it is often observed that there are as many interpretations as there are interpreters (or maybe even more!).

A further weakness in the "expert opinion" approach is that expert and nonexpert interpreters often disagree significantly in their interpretation of the theory they both claim to espouse. This is true as much in the case of science as in that of religion or ethics. Not only do experts and nonexperts typically disagree in their interpretation of a theoretical perspective or a tradition, but the former are notoriously inaccurate in their predictions of how the latter are likely to interpret it. Or, what is even more familiar, experts tend strongly to view the opinions of the nonexperts as uninformed and/or the product of irrational fears or influences (such as a sensationalistic media). Expert interpreters of a religion sometimes view the adherents to that religion as ignorant in their understanding of the theological or ethical tradition and thus may even question whether they "really are" adherents of the religion. This raises the age-old question of what (or who), finally, defines the nature of a religion: is it what the expert scholars of religion or its own theologians and interpreters say it is (or ought to be) or what the majority of those who claim to practice it say it is (or reflect in their own behavior)?

This book attempts to deal creatively with these problems of "expert interpretation" by engaging both the expert interpreters of these religious and ethical traditions and the "nonexpert" or lay adherents to the traditions in their understanding of the issues posed by GM foods to their dietary norms. The chapters that follow in the book are written by competent scholarly interpreters of the religious and ethical traditions we consider. But we have also engaged groups of lay adherents, committed practitioners of these religious and ethical traditions, in a process of informed conversation and debate around the implications of their dietary norms for certain specific aspects of genetic modification we felt they might consider relevant.

This more empirical and *descriptive* part of the research project has been carried out through the use of focus groups whose members were recruited from local religious communities in Western Canada.[2] These groups were conducted by a professional facilitator, using a carefully constructed set of questions relating to technological developments such as animal cloning techniques and, particularly, the transfer of genes from plant or animal sources considered to be unacceptable as food within their tradition. These groups were presented with a short presentation on the nature of GM food technology, designed to be as objective as possible, and then were asked to talk about their reactions to a series of different uses of this technology in light of their dietary philosophies and commitments. There is, of course, no assumption that focus groups are "representative samples" of a religious community. They clearly will not be. Their value is not that they provide a scientifically reliable picture of the attitudes of adherents of the tradition as a whole but that they help call attention to the ways these adherents can view an issue in light of their commitments to the tradition—ways that may well be missed, or even dismissed, by the expert interpreters of the tradition.

We have represented the views of the focus group members in this study within the chapters written by the scholars. An integral part of the process of this research project was two meetings of all the chapter authors and the focus group facilitator. At the first team meeting the authors agreed upon a methodology and "common focus" for the book as a whole and began the formulation of the questions for the focus groups. The second meeting of the team was held a year later, where the authors subjected their first drafts to the critique of the whole team and where they were presented with the results of the focus groups that had been conducted in the interim. Not surprisingly, the views expressed in the focus groups toward both GM food technology in general and specific applications of the technology often differed substantially from the interpretation of these same issues offered initially by the chapter authors. The authors were then asked to incorporate the views expressed in the focus groups into their revised chapters, not necessarily by *accepting* them but by trying at least to understand and explain the understanding of the religious or ethical tradition underlying the responses. As a result of this process of engagement with the focus group comments, and the peer review provided by the team meeting, nearly all the chapters express views on the genetic modification of foods that are supported both by the scholarly analysis of the tradition itself and by the understanding of this tradition expressed by lay members in the focus groups.[3] Further comment on the divergence of viewpoints between the focus groups and the chapter authors is contained in the final chapter.

Our hope is that the two-pronged methodological approach of this book provides a more profound understanding of the way in which adherents to these religious and ethical traditions, both expert and lay, are likely to exercise their deeply held value commitments around food production and diet in the marketplace and in the political sphere generally.

In the first chapter, Samuel Abraham sets the baseline for our consideration of "acceptable genes" by describing the scientific understanding of genes and how they work in genetically modified organisms (GMOs). A GMO is one where an alteration has been made to the genetic material that will be inherited by the offspring of the organism. Abraham explores how the current development of food and animal GMOs may be seen as an extension of a long-standing human effort to "improve" on desirable traits in agricultural plants and animals. After defining genes, Abraham goes on to describe their connection to simple and complex traits and how gene expression involves the production of proteins. He then explains how a transgene GMO (one in which species boundaries are crossed) is created and used in agricultural food production. For example, the genetic modification of plants can allow them to utilize less of an existing resource, to function better in a particular environment, or to provide extra value (e.g., more protein). In the future, says Abraham, the biotechnology industry will continue to seek new ways of genetically modifying existing agricultural practice so as to add value to the foods we consume for both the farmer and the consumer. The chapter concludes with a discussion of the risks involved in the application of genetic modification to the agricultural production of food—especially the lack of genetic diversity and genetic variation resulting from the use of fewer varieties.

In his chapter, Paul B. Thompson summarizes the ethical rationales behind using, opposing, or qualifying the use of agriculture as they have been expressed in the past twenty years of debate. He interprets the controversy over agricultural biotechnology as an episode in the ongoing social, political, and ethical struggle over the guidance of food production and consumption. He reviews this debate under five categories in which the products and processes of rDNA have been alleged to pose risk: human health (i.e., food safety), the environment, animal welfare, farming communities in the developed and developing worlds, and political and economic power relations (e.g., the rising importance of commercial interests and multinationals).

Having examined debates over the possible risks posed by gene technology, Thompson goes on to consider the use of gene technology as itself the basis of concern—that there is something about the manipulation of living matter at the genetic level that is of ethical concern. Some members of the general public, for example, hold the view that the manipulation of genes or cells is either forbidden or presumptively wrong (e.g., unnatural). This viewpoint was manifested in several of our focus groups. Arguments used to support the belief that the genetic modification of food is an unnatural activity include the following: (1) the idea that plants and animals have "essences" that are violated by genetic engineering; (2) that plants and animals are part of a "natural order" with an underlying system of specific purposes that genetic modification will disrupt; (3) that genetic modification of foods evokes a reaction of emotional and moral repugnance (e.g., the reaction expressed by many to the cloning of Dolly the sheep in 1996), which is viewed in itself as grounds to regard cloning as intrinsically wrong; and finally (4) religious arguments that attach religious significance to species boundaries and question the wisdom of genetic engineering, which crosses such boundaries (see chapters 5–11). As the chapters in this book show, although many philosophers and theologians reject religious and special concerns about the unnaturalness of biotechnology, most would support the argument that respect for such beliefs constitutes a powerful basis for segregating and labeling the products of biotechnology so that people can make choices.

In her carefully argued chapter Lyne Létourneau finds that genetic engineering does not pose a *direct* threat to vegetarianism. The case for vegetarianism finds no rational basis to reject the use of genetically modified plants with added DNA from animal origins. But there are *indirect* reasons for concern, and these have to do with an underlying conflict between genetic engineering and vegetarians' systems of values and beliefs about the social and physical environment. Létourneau distinguishes between "health vegetarians" and "ethical vegetarians." Health vegetarians adopt a vegetarian diet mainly for health reasons and are less committed ideologically to vegetarianism than ethical vegetarians. Consequently, health vegetarians are less likely to oppose GM foods—whether they include transgenes from animal origin or not—if the safety of such GM foods is clearly established and if they offer proven health benefits in addition. Health vegetarians constitute the majority of people who adopt a vegetarian diet.

However, ethical vegetarians base their opposition to the consumption of GM foods more in concerns over environmental protection and global considerations for social justice along with sensitivity to human health issues. Ethical vegetarians worry that the genetic modification of food will lead to a loss of genetic diversity; that it will widen the gap between developing countries and industrialized nations; that it will reinforce the concentration of power in the hands of industry; and that it will create unknown risks for human health. Even if human health benefits were established and proven to be risk free, the genetic modification of foods would continue to be opposed by ethical vegetarianism for social and environmental ideological reasons.

Létourneau notes that the focus groups involved in our study emphasized moral consideration for animals and environmental protection as a major basis for vegetarianism. Moral questions over the use of animals for food were a special concern in the vegetarian focus groups, specifically the view that it is morally wrong to treat animals as resources to be exploited or as a means to achieve human ends. Thus animal farming for the purpose of raising and killing animals is seen to be a violation of their moral status. The transfer of genetic material from an animal to a plant is described by one focus group member as a violation of the animal's essence and integrity. Thus the focus group's view that animals have intrinsic value also supports the position that there are species boundaries that must be respected and not crossed by genetic engineering. The conclusion from the focus groups was that adding animal DNA to plants violated the species boundary requirement; it incorporates into a plant an element of an animal's identity, and consequently eating such a plant may be considered as equivalent to eating animal flesh. For practicing vegetarians then, the labeling of such transgenic plants is clearly required.

Laurie Zoloth, a specialist in Jewish ethics, finds a surprising openness to GM foods and technologies in the Jewish tradition. The most respected legal authorities of the Conservative and Orthodox Jewish communities advocate the widest use of new ideas and technologies with reference to GM foods. This approach arises from an ethical perspective that begins by assuming that anything not listed in the Jewish texts as forbidden is permitted. Although the *Mishnah* prohibits the grafting of plants, this does not prohibit the joining of DNA strands from one species to the DNA of another species since DNA technology is not mentioned in the texts. Further, if something is not permitted to Jews to make or do, says Zoloth, it is not intrinsically immoral unless it violates the laws given to all peoples as Noah's family emerges from the ark. Thus, although certain sorts of cross-breeding or grafting are prohibited for a Jew, a Jew may still make use of the products of others' crossbreeding, for example, by riding a mule or eating broccoli. Other overriding aspects in relation to Jewish ethics and new innovation are the core values of saving life (*pekuach nefesh*) and healing the essential brokenness of the world (*tikkum olam*). In Jewish scriptures the world is seen as unfinished, and the role of a Jew is to be part of the ongoing act of creation that may be enacted through human interventions in illness, suffering, agriculture, and industry to better feed the world. These considerations, argues Zoloth, have led Jewish law to give a broad and sympathetic hearing to research in genetics and the genetic modification of foods. Yet Jewish ethics also contains general admonitions not to harm nature or to act in a hubristic fashion.

Zoloth helpfully describes the ethical understanding of the place of food in Jewish life. Nature is not seen as normative, and the production of food is instrumental. Food and the rules of its production are essentially the way God cares for the poor via our labor. The Jewish system of "kosher" is intended to place limits on the desire, production, and consumption of food. It is in the sharing of food at home and in the Jewish community that these nonrational rules of kashrut (kosher practice) still apply. With regard to genetically modified animals, grain, vegetables, fruit, and dairy products, Zoloth shows how contemporary Jewish legal scholars have studied the genetic modification of foods and carefully considered the Jewish prohibitions against mixing some animals and some plants. In most cases, it seems, genetic material may be transferred from one species to another without violating the prohibition against mixing. Commentators also turn to the possibilities for healing or for the better feeding of the poor in their support for GM foods. But worries are also expressed over the possibilities for unforeseen side effects. Zoloth notes these same worries as appearing in the Jewish focus group comments along with worries over the role of greed in the production and marketing of GM food. However, in spite of these worries, Zoloth, through a careful textual study, concludes that genetic engineering is not prohibited by Jewish law, and indeed may well be a way to help the world's poor to a good harvest of improved food leading to better health. However, in the focus groups of Jewish lay people, fears were expressed about the use of GM food from the perspective of its safety, its permissibility under the rules of kashrut, and whether the genetic modification of animals is an unacceptable act of human hubris in its alteration of God's creation. Zoloth tends to discount these worries by Vancouver-area lay people as a result of, as she puts it, their living remotely from centers of Jewish population and scholarship.

In his chapter, Donald Bruce of the Church of Scotland observes that the attitudes among Christians toward GM food vary widely from enthusiasm to outright opposition but often lie somewhere in between. Unlike many other religions, Christianity does not have specific food requirements. The *New Testament* declares that there are no prohibitions on any type of food. However, Christians are to be filled with the Spirit of God rather than engaging in drunkenness or excess eating. But sharing a meal is the ideal in Christian family life. As a member of the Mennonite focus group said, "we use food to show love in the act of eating together." Food also is used symbolically in the central Christian rituals of the Lord's Supper, Holy Communion, and the Mass.

With regard to the genetic modification of food, most churches that have examined the practice do not find a theological reason to say it is intrinsically wrong. Rather, Christian concerns tend to focus on the consequences or social context of such activity. The genetic modification of plants and animals to meet legitimate human needs is seen by many Christians to be acceptable as part of the dominion over nature granted to humans by God. However, humans are accountable to God in their actions, which must ensure respect for creation and love for nature and for the disadvantaged in the use of nature. Past selective breeding has greatly changed the genetic makeup of plants and animals. So, it is suggested, if God has given humans the skill to alter food crops and animals by moving a few genes across species, and there is no clear biblical prohibition, then the practice should not be rejected. God's creation may be seen as filled with possibilities that humans, through science and technology, have a God-given mandate to develop for the enrichment of the lives of others.

Some Christians, however, feel that genetic modifications of food and animals are not natural and are an attempt to play God by wrongly changing what God has created. This worry was expressed by some people in the Christian focus groups. Both sides of this debate within Christianity are carefully reviewed by Bruce. He observes that for many Christians the issue becomes a matter of judgment as to whether GM is more hubris or a right use of our God-given talents for human good. But what about questions of risk? Because we as humans have finite knowledge and are morally fallen in our understanding, we may not have wisdom to know the outcomes sufficiently to make such far-reaching changes as the transgenic modification of species. Some focus group members worried about the unknown effects that genes or proteins introduced into a plant from another species might have upon the body or on the rest of nature. Such worries lead to calls for the exercise of precaution, but how much? The Church of Scotland concluded that because the risks vary greatly between applications, a blanket moratorium against GM was inappropriate. Others in the focus groups argued that no genetically modified crops should be used until it can be demonstrated that no harm will result. Overall the strongest concern raised in Christian assessments of GM food is a concern over issues of justice and power. This is closely tied to questions of indigenous versus scientific knowledge (as discussed in the chapter by Shiri Pasternak, Nancy J. Turner, and Lorenzo Mazgul) and issues of justice and power in both developing and industrialized countries. For example, it is uncertain if many of the GM applications invented for North American bulk commodity farmers are relevant to the needs of African subsistence farmers and their equivalents around the world. Bruce also raises worries over the use of animal cloning in U.S. beef production as going too far into a factory mass production mentality and losing respect for animals as God's creatures. Some Christian laypeople expressed reactions of "abhorrence" to such procedures.

In his chapter on Muslim ethics, Ebrahim Moosa notes that the Qur'an views food consumption as part of the commandment to live a full life. Caring for the body is a key part of caring for the self and is essential for salvation. In this context, the Qur'an and prophetic reports identify foods that are banned from consumption but do not give reasons as to why pork and wine, for example, are prohibited. However, in the Qur'anic concept of '*fitra*' (the naturalness given by God in creation), even plants and animals have an innate disposition that determines their proper order. But *fitra* is subject to distortion through human sin and disobedience. How does this play out when humans get involved in altering plants and animals in agricultural science and practice? For Islam, the key guideline goes back to an encounter of Muhammad with workers out in the fields grafting different species of date-palm seedlings. Although at first Muhammad suggested it would be better not to graft, he later limited his authority to moral matters that explicitly impact on one's salvation. When it came to practical agricultural matters, he endorsed experience and expert opinion. In other words, ethical issues that are tied to secular or worldly pursuits and rely on scientific or empirical knowledge are to be decided on their scientific and practical merits. However, such ethical behavior in the secular realm is limited by our stewardship (*khilafa*), responsibility that we humans must exercise in our dealings with nature. Moosa concludes that Muhammad's example regarding grafting along with the questions as to what good stewardship may require leaves Muslim ethics ambivalent and undecided in its response to the challenge of genetically modified foods. It is at best a "work-in-progress" marked by its dearth of ethical deliberations apart from a few juridical responses or *fatwas*, which Moosa goes on to analyze.

For Islam, says Moosa, there is nothing that designates foods as good or bad, permissible or impermissible, in terms of their own inherent qualities. GMOs, however, because they are so unprecedented, lead most Muslim ethicists to view them as requiring a personal commitment to study and intellectual effort *(ijtihad)*. Over time, such accumulated effort will update the *shari'a* or ethical canon so as to be able to respond to the challenges of GM foods. At present Moosa observes two trends on the issue of GMOs: (1) the group of traditional religious authorities (along with more technocratic and professional Muslims) who give ethical and legal support to GMOs while viewing them as manageable risks; and (2) Muslim professionals and technocrats who discuss GM foods in terms of a precautionary approach. An example of the first trend is the Saudi-based Council for Islamic Jurisprudence (CIJ), which has studied GMOs since 1998 and ruled that it is permissible to employ genetic engineering in the sphere of agriculture and animal husbandry so long as precautions are taken to prevent harm to humans, animals, and the environment. But little guidance is given as to how such harm is to be identified and measured. The same CIJ ruling, however, did insist that the use of GMOs in food be disclosed through labeling. Muslim authorities in Indonesia, Singapore, Malaysia, and India have also given a cautionary green light to GMOs in the human food chain. In North America, the Islam Food and Nutrition Council (IFANCA), which designates foods as permissible (*halal*), is reported to support foods derived from GMOs.

Regarding the second trend, "the precautionary approach," Moosa points to Muslim communities, especially in the West, where it is modern educated Muslims with scientific training, rather than religious authorities, who voice reservations about GMOs in foods. With the help of some Qur'anic knowledge and a little Islamic theology, such individuals adopt a more critical stance toward genetic science and agricultural biotechnology. Here Moosa points to Mohammed Aslam Parvaiz, who aligns himself with those in the scientific community who believe that the use of transgenes in food harbors catastrophic environmental consequences. Parvaiz sees GMOs as janus-faced, producing both innovations and disrupting disturbances in Allah's creation. Parvaiz, in his resistance to GMOs, resorts to a theological reading of Qur'anic passages that urges humans not to alter God's creation. Moosa offers a critical assessment of the interpretation. As an example of a different sort of precautionary approach, Moosa discusses Saeed Khan, who is less theological in his urging of Muslims to join the alliance of concerned scientists, producers, and consumers in the United States and abroad to combat the use of GMOs in food. Khan sees GMOs as an alarming use of science to potentially colonize people in both the developed and the developing worlds. Canvassing opinions from the Muslim focus group, Moosa finds views that fit with both the managed risks and the precautionary approaches outlined above. Overall, the lay members in Muslim focus groups showed hesitation and ambivalence toward embracing genetically engineered foods and insisted that such foods be labeled so that Muslim consumers could avoid them. The use of a pig gene to enhance tomatoes met with strong disapproval, since pork is unlawful for Muslims. Others in the focus groups argued that genetic engineering would upset the natural balance of nature, especially when genes from one species are mixed with another. However, some discussants were favorably disposed to genetic modification and suggested that experts in Muslim law should make the final decision on such questions. Moosa notes that some focus group members, like some scholars, were worried by long-term risks from GM food, as yet unknown. Moosa also observes that Muslim scholars have not taken the views of Muslim laypeople into account in their responses to the genetic engineering of food. Moosa himself sides with the precautionary approach.

Vasudha Narayanan, in her carefully nuanced chapter on Hindu attitudes to genetically modified foods, makes clear that food is a big topic in Hindu dharma or law texts. Hinduism is a diversity of communities with various castes, philosophies, and geographical areas, all of which bear on food. Some Brahmins, for example, are strict vegetarians, but others may eat fish. If they are followers of Lord Vishnu, they will not only be vegetarian but will also refuse garlic or onion. Food is also of central importance in religious ritual activities where it is classified as pure or impure. While impure food may be eaten on a daily basis, food taken on certain holy days

or used as an offering to the deity should only be pure. It is this distinction that applies most directly to GM foods that are judged by Hindus to
be impure. Discussion in the Hindu focus groups reflected this sophisticated
approach to food. This is especially the case for transgenic foods in which
an animal gene has been introduced into a vegetable or fruit. Such a food
would be considered to be impure and not suitable for use on ritual days
or as an offering in worship (*prasada*). However, such GM foods, although
impure, could be used on a daily basis if there was no health hazard and if
the food alternatives available would be worse in the minds of the Hindu
devotees. Thus, if the cheeses available were made with animal-derived rennet and other artificial substances, they might be accepted by Hindus for
ordinary but not ritual use. Overall, for Hindus, the origin of the gene is
as important as the traits it may have associated with it. For example, if
a gene originated from a pig it may not be acceptable to many Hindus as
demonstrated in the opinions expressed in the Hindu focus group. While
Hindus' attitudes toward the daily use of food are quite diverse and open
to new developments (including even genetic modification), when it comes
to religious rituals or use of food in worship only pure food (not foreign or
GM) may be used, and this strict attitude does not change.

In his chapter, David R. Loy shows that although Buddhist ideas arise
in a Hindu context, the Buddhist approach to food and its genetic modification is quite different. Rather than focusing on the ritual purity or impurity
of the food, Buddhism shifts the spotlight to the motivations behind our
use of food and our institutional or collective reasons for its genetic modification. Relating his analysis to current ethical theories, Loy shows how
Buddhism adopts a "virtue" rather than a utilitarian or deontological basis
in its approach to food and the acceptability or not of its genetic modification. After outlining the traditional Theravada and Mahayana Buddhist
approaches to food, Loy shows that while vegetarianism is emphasized in
Mahayana traditions (especially in China), the key thing for all Buddhists is
the intentional motivation involved in individual, collective, and corporate
choices and whether this increases or reduces *dukkha* (suffering) for us as
individuals and for the ecosystem (animals and plants, earth, air, and water)
in which we humans are but an interdependent part. As a participant in the
Theravada focus group put it, taking a gene from one species and transferring it to another might be acceptable as long as "it is going to improve
the food and if it is for the good of the whole world." Or, in the words of
another focus group member, "It is not just about scientific capability but
whether we should do it." But lack of labeling violates the right of choice
needed for one's Buddhist practice.

Loy shows how from a Buddhist perspective our motivations and
choices construct who we are as persons and how we can reduce our *dukkha* or suffering by transforming the three unwholesome roots of human

motivation, namely, greed into generosity, ill will into friendliness, and the selfish delusion of a separate self into the recognition of being an interdependent part of the world's ecosystem. Unlike Hinduism, the focus is more on becoming a different kind of person in this life than on the possibilities of rebirth with its emphasis on *anicca* or the idea that everything (including ourselves) is constantly arising, changing, and passing away. Buddhism is open to new technologies and progress. For Buddhism, technologies such as the genetic modification of food are not a problem in and of themselves; it is the motivations behind such modifications, our use of them, and the effects on our *dukkha* (suffering or happiness) that is important. Does GM food increase or decrease our *dukkha*? The Buddhist answer involves seeing that everything is interconnected, both natural phenomena and our human technology. Thus in evaluating how GM food may increase or reduce our *dukkha*, it is often unexpected side effects that are important, for example, Bt corn pollen seeming to be poisonous to monarch butterflies.

What do Buddhist principles imply about GM food? Loy says that the three unwholesome roots of motivation (greed, ill will, and delusion) must be extended from the individual level to how they operate collectively and institutionally in the food industry. Here the current difficulties in testing for adverse side effects, along with corporate pressures for short-term profit and rapid growth, pose the following question: can the food industry subordinate its own interests in GM and place top priority on safeguarding the needs of human consumers and of the whole ecosystem? Loy's Buddhist analysis and the views of the Buddhist focus group put the ethical focus on "institutional greed" and "institutional delusion" (forgetting that we humans with our science and technology are but a part of the biosphere and its ecosystem). As Loy puts it, there are no side effects. Since we are part of the natural world, if we make nature sick, we become sick, and our *dukkha* increases. This is how karma operates.

Loy concludes that the genetic modification of food as currently practiced seems incompatible with the kinds of personal and collective motivations necessary for both human and ecosystem *dukkha* to be reduced. A fuller understanding of the genomes of plants, animals, and humans and how they can affect each other is needed if our ambitions or greed are not to outrun our wisdom. However, GM food is not necessarily always a bad thing—some future types of GM modifications (e.g., vitamin A enriched rice) might serve to reduce some types of *dukkha* in our world. But for this to happen, GM technologies must be evaluated from within the larger social, economic, and ecological contexts within which they are devised and applied. Here, the Buddhist approach requires that the personal and institutional motivations be seen as a key part of that context and be given a central position in any evaluation.

In her chapter on the Chinese approach to food and its genetic modification the historian Hsiung Ping-chen finds the basic principle for food to be so that "you may have it with no harm." Historically the Chinese approach has a variety of food traditions in which there are in general no religious taboos. Instead the Chinese focus is on achieving balance between *yin* (cool) and *yang* (hot) foods so as to foster both health and taste. There is no separation between food and medicine; medical effects are expected from ordinary food such as congee, or rice porridge. Hsiung shows how the eighteenth-century author Ts'ao T'ing-tung in his *Book of Congee* maintains that although congees are also a medicine, comfort and taste are of ultimate importance. So the author divided congees into the categories of superior, average, or lowest depending on taste and smell. In the preparation of congee the ingredients used (e.g., rice, water, and fire or heat) are of crucial importance so that the *yin-yang* balance can be maintained along with good taste. All of this is crucial for the health of people especially as they become old and frail.

According to *The Book of Congee* the most superior congees are made with vegetarian ingredients flavored with mild-tasting animal products such as cow's milk, chicken broth, or duck broth. Lesser congees were cooked with stronger tasting meats from deer, sheep, pigs, or dogs. This traditional Chinese approach to food, while open to change, regards genetic modification of food as human tampering with nature and something thus to be avoided. Worries are voiced over the unknown effects genetic modification of food can have on human health and the environment. For many Chinese, the genetic modification of food is seen as a violation of nature and its balance—essential for human health. Thus, food labeling is needed so that individual choice is possible.

Turning to indigenous peoples we find that they do not share a common religion, but most share a common history of colonization and Christian missionization. In their chapter, Pasternak, Turner, and Mazgul argue that for indigenous peoples genetically engineered foods are an extension of the ongoing worldwide colonial destruction and desecration of their local knowledge systems and the natural world. Indigenous cultural practices relating to food production and consumption have been central to preserving and transmitting to future generations local ecological knowledge, social institutions, ethnic identity, and spiritual teachings. A common position shared by indigenous peoples around the world is the essential connection between traditional foods and practices for the maintenance of their cultures—a connection currently being challenged by the genetic modification of food. This chapter offers two case studies into this situation. In the first, Mazgul and Pasternak traveled to Guatemala to conduct focus groups with the Maya community. In the second, Turner sent out questionnaires to indigenous colleagues in North America with a specific interest in food. In both case studies the focus was on the impact of transgenes and GMOs upon traditional food-gathering practices and the belief systems in which they are embedded.

In the Maya case study, corn is seen to be not only an essential traditional food but also crucial to the Maya spiritual worldview and ecological practice. In focus groups, corn is described by the Maya participants as "our mother," and members talk about rituals around planting, harvesting, and eating that are being lost under the impact of colonial and corporate food practices. Similarly, all of the North American indigenous questionnaire respondents emphasized the essential connection between traditional foods and cultural practices and how this connection is under threat from the development of biotechnology, of GM foods, and of the global market economy. On this point there is no divergence between expert opinion and the Maya focus group. Also, the privileged position given to science in modern life is questioned in relation to traditional wisdom regarding food. Rather, science should be seen as one story among others and not as a hegemonic truth that trumps traditional food knowledge systems.

In these case studies the majority of indigenous participants rejected outright the eating of GM foods under any circumstances. The introduction of DNA or genetic material from culturally prohibited foods to acceptable foods through genetic engineering techniques was judged to be offensive—a violation of spirituality, of cultural practice, and of the natural order. This violation was imposed by colonial forces, capitalism, and corporate power. Some even described it as yet another attempt at the cultural genocide of indigenous peoples. Another point is that indigenous peoples generally view plants and animals as different clans of people. Thus the question of genetically modifying food is of the same order as genetically modifying humans. However, some Mayas did say that they might tolerate small modifications to domesticated animals such as pigs. But if it is a sacred plant such as corn, no genetic modifications would be tolerated. Also, in the Maya view God created animals and plants that in themselves are perfect. Therefore, the idea that genetic modification can improve on them is unacceptable. It is important to know by labeling if something is a GM food so that it can be avoided for cultural and spiritual reasons or because there may be unknown long-term health risks.

In the concluding chapter Conrad G. Brunk, Nola M. Ries, and Leslie C. Rodgers consider some of the regulatory and market implications of the above chapters. They pick up the worries over GM and transgenic foods expressed in the preceding chapters and the need for labeling that results. They explore the policies that need to be put in place to ensure that the labeling of food for human consumption will allow individuals to make the ethical and religious choices required for the practice of their own beliefs.

While these chapters contain much diversity in their views on the position of animals and the acceptability of the genetic engineering of food, one point stands out as held by all traditions, and that is a focus on the underlying motivation. If animal genetic modification is meeting a real human need, it may be seen as acceptable. However, if it primarily reflects individual or corporate greed, or a scientific drive to be first, then it is not considered acceptable by any religion. Also, the possibility that the natural-ness or species integrity of animals may be challenged by genetic engineering is a concern adopted by some theologians and by the focus groups of all traditions. Doing such things as cloning or transgenesis to animals created by God generated a sense of abhorrence among laypeople in virtually all focus groups. This strong sense of abhorrence is not always shared by the expert opinion of theologians. Thus, in some traditions a conundrum results on this point between the views of ethics theologians and the laypeople of those traditions.

Notes

1. Mainstream Christianity, as noted in the chapter by Bruce in this volume, does not generally have any prohibitions on any food sources, but Christianity does have norms relative to the just production of food, as exhibited in the Christian focus groups.

2. The exception was the aboriginal focus group conducted in Guatemala within a community with whom one of the chapter's coauthors had family and tribal connections.

3. One exception to this is the chapter by Laurie Zoloth on Judaism, which is more critical of the opinions expressed by lay Jews in the focus groups.

I

Genetics and Genetically Modified Organisms

Samuel Abraham

The intent of this chapter is to provide the reader with an understanding of genes and their contribution to the inheritance of the traits they influence. In describing genetically modified organisms (GMOs) a number of scientific and practical definitions of genes, how they work, and how they can be manipulated will be explored. Throughout this chapter I shall also examine some of the drivers to the development of GMOs. I shall attempt to paint a picture of how our current foray into GMOs might be seen as an extension of a never-ceasing march to "improve" on desirable traits that goes back to our roots in agriculture.

A GMO is one where an alteration has been made to the material that will be inherited by the offspring of the organism. Such alterations can be in the form of additions to, subtractions from, or substitutions to the existing material passed on. This material, known as deoxyribonucleic acid (DNA), is common to all known living things and is the means by which the specific characteristics of a parent organism are transmitted to its offspring. The changes to the DNA will typically result in changes to the inherited characteristics of the organism.

The Gene: A Simple Definition

A "gene" is often described as the string of DNA that contains all the necessary information for the making of a specific protein, the machinery in cells. DNA is a long string of four repeating molecular building blocks that are brought together in different combinations to be the code of all genes. It is the "reading" of the different codes (arrangements of the four molecular building blocks) that allows for a gene to be expressed into protein. This reading is the process by which a cell copies and then translates the code into the protein spelled out by the code. The protein in turn acts to fulfill a specific function in the cell or organism in which it is being expressed. The length of DNA and the code for a gene are specific to the function of the protein encoded for. The genes of all living things, as we know them, are comprised of the same basic molecular building blocks of DNA. In fact some of the same genes found in one living thing (e.g., yeast) are also found in other vastly different life forms such as humans, usually because they are part of biological processes that are key to such basics of life as using energy derived from external nutrients. Variations in the code of a gene within a single species are called "alleles." An allele is simply a variant of a gene that may or may not result in functional differences to the protein produced.

The Gene: A More Complex Definition

In truth however, a gene is far more complex than allowed for by this narrow definition. To begin with, in addition to the structural information for creating the protein, all gene expression is tightly controlled by one or more regulatory sequence(s) that are a part of the overall stretch of DNA that encompasses a gene. Gene expression is the control of when, where, and how much of a protein is produced by its particular gene or code. The DNA comprising a gene by this view can therefore be broken down into two parts: the parts that have the information for the making of the protein itself, the structural domains, and those segments that control for gene expression, the regulatory domains. Alleles of genes can occur by variation in either the structural or regulatory domains of a gene. The result can be a protein that varies by its expression level or else in the functioning of the protein variant for which it codes. Regulatory domains can be found at several places along the DNA depending on the gene. Structural domains are often interspersed among the regulatory sequence and can be read selectively resulting in proteins with different roles being produced from the same gene. By this selective reading, a single gene can produce several related proteins that differ by sections of code in the gene that are brought together in a manner analogous to the copying, cutting, and pasting of sentences or words in word processing software, resulting in different paragraphs with related but not identical information.

In summary, while genes are often presented as simple linear units of genetic material (DNA) that are read from one end to another, their regulation, and the fact that they can often produce variations on a theme (as to the protein produced), means that the full story is more complex. Further to this point, the regulatory sequence encoded in a gene regulates gene activity through the binding of specialized proteins to these sequences. The specialized proteins are coded for by their own unique DNA code and in turn have their expression regulated via their regulatory sequence by yet other proteins. By the extension of this scenario virtually all genes in an organism are now inextricably linked to each other's expression. This then brings us to the concept of the genome, the entire repertoire of genes needed to code for any living thing. This view of the interrelatedness of genes and the genome in the context of an organism might be useful in considering the effect of the changes wrought in the genetic modification of plants and animals.

Humans have an estimated thirty thousand genes[1] that are differentially controlled in time (during our development both in utero and as we mature exutero) and space (in the different organs and tissues that comprise us). Through this complex and well-orchestrated expression of tens upon thousands of genes and their differential expression, cells, organs, and various tissues develop and maintain their identity and integrity.

From Genes to Traits

A trait is any quantity we can measure or observe, from height and weight to the amount of protein to be found in milk or else a measure of a behavior. The study of traits in all living things examines the effect of a gene or genes in combination with the influences of environmental factors (level of nutrition or prevalence of disease, etc.) in yielding the final trait that we observe. The majority of traits are complex as they involve many tens if not hundreds of genes acting in concert with the environment to affect the final trait witnessed. For example, in the study of human disease it is recognized that the majority[2] of diseases owe their clinical outcomes (measurable traits of different endpoints in clinical disease) to the consequence of the many different alleles we inherit acting together to result in different outcomes or end points. So, while a few genes might be involved in causing the disease, patient outcomes are affected by the complex interactions of other inherited alleles of different genes particular to the individual. For example, the disease ascribed to changes to the Alpha-1 gene[3] manifests a range of outcomes from subclinical (no effect on an individual carrying the abnormal version of this gene) to severe (death due to loss of lung function in the adult).

To determine whether a trait is simple or complex, the impact of the expression of a single gene on a trait can be described by a simple equation where a value (high or low) can be ascribed to the gene. A gene with a high value codes for a simple trait where other genes and environmental factors matter little in the equation, and therefore the full manifestation of the trait would always accompany the inheritance of such a gene. A gene with a low value has a much less predictable impact on the final manifestation of the trait. In this case numerous genes all contributing a small part to the overall equation in addition to environmental factors (e.g., smoking and susceptibility to lung cancer) would represent a complex trait. Simple traits are typically traits that are highly reproducible and more likely to be stably inherited and by and large are the traits involved in GMOs to date.

Genes as Scientific Tools

The science of manipulating a genetic code in structural or regulatory domains has greatly facilitated our understanding of the contribution a particular gene makes to a trait being studied. When studying a gene and its function(s), scientists have classically resorted to using cells grown in plastic containers in the laboratory. The cells in question are derived from many organisms such as plants, insects, animals, or microbes and are grown under specific conditions allowing for their continued propagation. Such studies frequently look to overexpress a gene and therefore the resultant protein by changes in the regulatory domain of the gene in question. The effect of an excess of the protein is observed by looking for an alteration in a measurable trait in the functioning or fate of the cells where it is being overexpressed. Alternatively, the structural domain(s) of a gene can be altered and then the cells observed to determine the effects of the modified proteins. This powerful process has brought many potential gene/protein functions to light.

Additional experimentation often involves looking at the altered expression or altered structure conferred upon a protein in living organisms (animals or plants). Such gene-expression studies look at expression of a gene in three possible ways: ubiquitous expression, meaning production of the specific protein in any and all tissue in the body at all times; expression in a tissue-specific fashion such as expression only in the liver; and expression in a developmental specific fashion, such as expression only during the time of ripening in fruit. Specific direction of desired patterns of gene expression is orchestrated using different regulatory sequences from genes already known to be differentially regulated in different parts or developmental stages of an organism.

These studies have allowed for a determination of both the potential broad and narrow influences of a specific gene on the function and fate of different cell types as involved in tissue and organ development and integrity (again a measure of traits). In summary, changing a gene in either the structural or regulatory domains, and observing the resulting change to a measurable trait either in whole organisms (plants or animals) or cultures of cells became a standard scientific technique for studying the functions of specific proteins.

While changes to the structural domains are instructive as to the manner by which the protein coded for works (i.e., the trait affected), experimentation with regulatory domains has been particularly important in enabling the creation of GMOs. Expressing a gene in an organism is best accomplished when the switches (regulatory domains) that allow for its expression are readily recognized and turned on by appropriate proteins already resident in the particular organism. This is more easily achieved when the regulatory domains used are derived from the recipient or target organism itself or from other closely related organisms. A transgene GMO is then created by first marrying a suitable regulatory domain to a structural domain encoding a protein of choice, then introducing this into the organism of choice to result in the stable, tailored expression of the protein in the GMO and its offspring.

Genes, Traits, Foods as Commodities

The fact that the genes of all living things are defined by the DNA they are comprised of and the recognition that some genes (as defined by their structural domains and their resultant proteins), if expressed in large enough quantities, might yield beneficial, marketable products led to early efforts to produce these proteins using cells as factories that were not necessarily of the same origin as the gene/protein from which the "desirable" gene was derived. For example, yeast or bacterial cells were used to produce valuable mammalian proteins such as insulin.[4] The reason for this is often simply the ease with which certain types of cells can be grown in large amounts to result in good yields of the protein being expressed.

All manner of cell types have been used to produce proteins of interest that are then isolated and utilized as needed. The juxtaposing of specific regulatory domains derived from the host organism next to the structural domain of any foreign gene is key to producing stable transgene GMOs with heritable traits. This stable expression of a foreign gene is achieved when proteins indigenous to cells of the recipient organism recognize and bind to the regulatory sequence that they normally bind to in bringing about the expression of structural gene(s) native to the recipient organism. This ability to express a foreign gene in another organism can then be taken a step further by tailoring the expression of a gene to specific tissue within the organism (e.g., the expression of a gene that impedes cell growth and maturation in the anther of corn so as to guarantee male sterility) by the type of regulatory sequence used in the making of the transgene.

A powerful consequence of these studies was the recognition that genes with what are potentially commercially useful effects could be expressed in organisms other than that from which they are derived in order to achieve a benefit hitherto unknown to the organism into which it is placed (e.g., the bacterial gene that confers resistance to the herbicide will allow for a commercially valuable plant now expressing a transgene derived from the bacterial gene to grow under conditions where the herbicide is used to control weeds).

The majority of current, commercially propagated GMOs are the result of either introducing a single "foreign" gene into a recipient organism or the modification of an existing gene(s) in the genetic makeup of an organism. In both cases the aim is to have these genes passed on through the germ line (those cells in an organism that allow for the gene and therefore the trait to be inherited by all the offspring with every passing generation). This permits a stable inheritance of the "transferred trait," the measure of the gene's effect as conferred by the protein encoded in this GMO.

A subset of GMOs, foods, mostly plants as opposed to animals (GM foods), have seen by far the majority of developments in this arena of biotechnology. Rapid and innovative developments that limit gene expression both in time (i.e., when in the development process of the organism) and space (i.e., where in the organism) have seen the clever use of regulatory domains in combination with structural domains encoding for simple traits being exploited with the intention of benefiting both farmer and consumer with respect to cost of food production, nutritional quality, and even the abrogation of environmental impacts / contamination.

Whether the consequence of expressing a foreign protein by utilizing native regulatory domains will cause a measurable change to the organism's full range of functions and responses is a key question for many. The act of insertion of a transgene into the recipient organism's DNA to allow for its stable expression and inheritance can potentially disrupt an existing gene in the DNA of the recipient organism. Newer laboratory techniques that allow a precise targeting in the insertion of foreign DNA into a host organism genome (homologous recombination) might alleviate this as a concern. The likelihood of increased metabolic demand in producing a foreign protein is likely to be negligible under standard management practices, while the likelihood that a gene that is key to the organism that has been disrupted is probably minimal if growth and maturation of the GMO remain unaffected. A GM food crop would obviously not be pursued if it did not have satisfactory growth characteristics, whatever other advantages it might have.

While growth and production can be positively influenced by a transgene, the possibility of negative impact to human health is typically examined by regulators only from the point of view of the safety (toxicity, allergenicity) and nutrient data pertaining to the particular GM food versus its non-GM counterpart. A GM food that is determined to be "substantially equivalent" to the non-GM variety in terms of such factors as composition, effects of processing and cooking, toxicity, and allergenicity is considered to be safe, as described by the Food and Agricultural Organization (FAO, established in 1945 as an independent arm of the United Nations) in a 2001 document titled "Genetically Modified Organisms, Food Safety and the Environment."[5] Like government agencies in many countries, Health Canada in turn has its own guidelines for safety assessment of novel foods.[6] When examined, scores of GM foods have been granted "no objection" status based on findings drawn from these guidelines.[7] While the purpose of the guidelines is to ensure that the GM foods give us no less than their non-GM counterparts and certainly pose no adverse issues, they also gauge the well-being of the plant from its general growth and production characteristics under standard management practices. In all, the guidelines have a very utilitarian, human-centric perspective with the plants only considered in the context of how well they suit our needs.

From Important Traits to the Green Revolution: A Continuous Enhancement of Traits through Genetic Modification of Agricultural Products

The selection of plants and animals for domestication and food production goes back thousands of years in human history. The traits that caused certain plants and animals to be chosen over others in a particular environment had often to do with key traits such as self-fertilization (for reproducibility of a trait—"breeding true"), time to maturation, yield (protein, carbohydrate, etc.) where plants were concerned, and additional issues such as degree of ferociousness and biddability as far as animals were concerned. The ability to improve the production of food yielded through better management of crops and animals has been central to the success of many larger societies in history as well as in present day politics and nationhood.[8] The development of agriculture, a new ability to support a large population, and the subsequent demands for greater production involved an ongoing interplay of population size and technological advancement, where one component reaches a threshold before the other can make rapid advancement.

The Green Revolution of the midtwentieth century saw the boosting of crop yields by the industrialization of agriculture through mechanization, irrigation, and the use of chemicals for fertilization and pest control. The yield of corn from an acre of Iowa farmland went from 26 bushels in 1929 to more than 130 bushels,[9] and countries such as India went from impoverishment due to an inability to feed their citizens to having had for many decades a surplus of food production available for export.[10] While the selecting of crops for desirable genetically determined traits (as defined by common sense and commercial parameters: yield, generation time, drought tolerance, disease resistance) is a perennial practice of farmers everywhere, the Green Revolution through the manifestation of large industrially run farms brought such selection to greater focus by further control of the environment through various management practices. Food crops that respond well to these management practices and produce high yields as defined by tonnage per acre are then adopted by a wider number of farmers in order to compete in the marketplace (in this case the selection pressure is two-fold: on the farmers to compete for higher yields and lower production costs and on the crops that best respond to the management practice). An outcome of this type of practice has been to focus on deriving seed for every succeeding generation that is drawn from the top performers of the previous generation. Competition among farmers and societal pressure for cheaper foods then result in a rapid adoption of those plants and animals with the most commercially viable traits.

Methods for selection of such desirable traits in plants and animals have traditionally involved observing the trait in the particular plant or animal and then using its seed to give rise to the next generation. The net benefit of this has been to produce a greater measure of the trait in every subsequent generation to the benefit of the farmer and society. As an example of traditional animal husbandry methods, in dairy cattle, such selection is done by noting the trait (e.g., milk protein production) in the daughters of several bulls and then utilizing the semen from the bull with the best overall milk producing daughters to give rise to the next generation of cows. Such techniques are particularly fruitful in dealing with complex traits where many genes are responsible for the trait under consideration. A careful breeding program, which then monitors the traits under selection both in the parents and their offspring, will ensure that the same alleles responsible for the traits in question are maintained in the population. In plants, traits ranging from early ripening to greater yields have also led to

those seeds in crops that self-fertilize being used into the subsequent generations. Self-fertilization, a trait common to most plant crops, allows a trait to
breed true (to be passed on fastidiously from one generation to the next).

In all, continued selection has helped lower the cost of food production and has therefore arguably made foods more available to more people.
The same promise is held out in the debate on GMOs. The promise of better yields, lower cost of production, improved nutritional value, and other
commercially attractive traits is often touted by the vendors of GMOs and
serves as a powerful incentive to farmers to adopt these new innovations to
improve their own competitive edge in the marketplace.

A consequence of selection for particular traits in modern day agriculture is that only a small proportion of the population (a few individuals: the
cream of the crop) is chosen to give rise to subsequent generations, which
results in a significant decrease over time in the genetic variation to be
found in food crops and animals under such selection. For example, in the
United States there were only 52 varieties of corn under cultivation in 1983,
as opposed to 786 varieties in 1903.[11] This significantly reduces the genetic
variation (number of different alleles for any one gene) normally to be found
in a population. This loss of genetic variation would likely represent a problem were selection pressures to change significantly (e.g., disease prevalence,
climate change, etc.), as the availability of fewer variants would reduce the
likelihood that there would be one or more able to cope with the changes
and allow for the survival of the plant or animal population in question. This
process is further exaggerated where the adoption of GMOs is concerned due
to the accelerated rate at which new strains can be generated, as well as the
rate of uptake of these new strains due to the touted benefit(s).

While most agriculturally important plants arose from one if not a
few individual ancestral plants, their subsequent spread across the globe has
resulted in the emergence of spontaneous genetic variants and the selection for those variants best suited to the environments in which they have
been husbanded. It is this genetic diversity, reflecting a historical journey in
geographic as well as cultural terms, that is often assailed by modern agricultural practice. GMOs will further the rate at which this will occur and are
therefore more likely than not to result in an even more greatly exaggerated
genetic constriction of crop plants. While biodiversity was essential in the
propagation of both plants and animals to suit their environments, it has
become less of an issue in present day agricultural practice due to stringent
management practices such as intensive irrigation, application of fertilizers,
and control of ambient conditions by use of greenhouses. However, biodiversity is still important in much of the developing world as the ability to
control production through such extensive management practices remains
beyond most farmers' means.

GM Foods

Several decades of reaping the benefits of increased food production made manifest by the green revolution and the burden of a burgeoning human population have created the habit of looking to the next technological innovation in order to improve yet further the traits we look for in foods (e.g., cost of production, hardiness, yield, value-added nutrients). The creation of genetic hybrids and techniques that cause multiples of the normal genetic content to occur have yielded products that have more robust characteristics overall as well as increased yields. All these techniques have profoundly changed the genetics of the parent varieties from which they were derived. While all such organisms are undoubtedly genetically modified, none is regarded as a GMO.

The label of GMO is usually reserved for organisms developed by combining certain techniques of cell manipulations and transgene transfer. A variety of techniques are used to effect gene transfer such as the use of an electrical current to get transgenes into a cell, or the use of bacterial or biological agents (e.g., Agrobacterium tumefaciens) or gene guns that literally shoot microscopic pellets coated with the transgene DNA into plants. Another group of techniques used to create GMOs consists of those where the genetic modification involves creating and screening for changes to genes already resident in the organism being manipulated. Such techniques use chemicals or radiation to induce random mutations in plant cells, which are then screened for protein changes relevant to the trait sought. Organisms produced this way tend to escape public notice as GMOs as they do not owe their newly acquired trait to the addition of "foreign" DNA.

While conventional plant breeding and selection practices have often focused on complex traits (e.g., yield, time to maturation), the nature of the techniques involved in the generation of GMOs means they are more typically focused on one gene and its influence on simple traits (e.g., herbicide resistance as might be conferred by a single protein as coded for by a single gene). The above techniques have enabled a more rapid turnaround in both the generating of new strains and the selection for the best among those conferred with the new trait. In many cases techniques that allow for detection or measurement of the protein being sought via the transgene allow for the rapid selection of cells to be used in the making of the final GMO without having to grow a plant first to achieve the full expression of the trait desired. This dramatic reduction in the time taken to create new strains of plants carrying genes for any simple trait with potential utility has made for very rapid dissemination and adoption of GMOs. Thus a fundamental difference between the current generation of GMOs and conventional breeding of plants or animals is therefore both the nature of the traits (i.e., simple versus complex) and the rate at which new strains can be introduced due to the techniques used to generate them. Finally, this technology permits the transfer of genes between species, and even between animal and plant kingdoms, that would not be possible with standard breeding techniques, since the donor and recipient species cannot breed naturally. The combination of these capabilities is what gives GMO technology its tremendous power, that is, the rapid derivation of a new trait and from potentially any living source. It is this very potency that then raises questions about the potential health and environmental impacts of these changes as well as the ethical implications.

The majority of plants targeted for development as GM foods are the traditional plants (maize, canola, wheat, soya, peas, etc.) that have had their traits selected upon for thousands of years. Transgene GM foods have as their main innovations the prospect of boosting or modifying existing traits in foods (e.g., higher protein content in grains; pesticide/herbicide resistance) or adding traits not traditionally found in a particular plant (e.g., beta-carotene in rice, herbicide resistance). In the repertoire of GM foods, plants have to date outstripped similar developments in animals. One reason is the nature of self-fertilization among many agricultural crops (a trait that was desirable in their ability to breed true, the very reason the original ancestral plant was selected for; unlike animals, which of course do not self-fertilize). Next is the ease of obtaining seed from relatively few plants to generate all that contribute to the next generation, with very little cost associated outside of distribution.

As mentioned above, strategies involved in creating GM foods from plants range from using chemicals or radiation to causing random genetic changes in the DNA code (across the entire genome) that are then selected for specific desired traits (e.g., resistance to herbicide), to actually adding a gene (transgene) from another organism that confers the desired trait (e.g., Bt toxin from bacteria into plants to confer pest resistance). The latter technique involves putting together a gene with both structural and regulatory sequences into cells that will grow into a mature plant and looking to see if the seed from that plant passes on the trait to all its progeny, while the former typically examines thousands of cells and selects for those that have undergone changes to the structural regions of particular existing genes by screening for the trait being sought before causing the right cells to become whole plants. In the debate concerning GM food it is the crops that result from the addition of foreign DNA (the transgene) that have typically borne the brunt of criticism when labels such as "frankenfoods" are used. Those plant crops that result from random gene-alteration (mutagenic) procedures followed by screening for the desired trait are not seen in the same light. One perception is that the conferring of a trait involving the addition of unrelated genetic material (transgene) not otherwise found in an organism is abnormal and might engender toxicities and allergenicity not previously known in that food. However, it is worth considering that the methods involving random mutagenesis through the use of chemicals or radiation are far more likely to cause other unknown genetic changes due to the indiscriminate nature of the procedure(s). Without knowledge of the changes to the DNA that might occur, these changes cannot be functionally screened for and will consequently be "silently" inherited by subsequent generations.

If a particular GM food reduces cost of production, helps improve yields, and is seen to include traits (e.g., beta-carotene) that are beneficial and therefore desirable for public consumption, its adoption under current farming practice would be (and has been) rapid. The worldwide arena of approved GM crops under cultivation has grown steadily since 1995, increasing by 15 percent in 2003 and 20 percent in 2004.[12] The practices of modern agriculture, such as the buying of seed from a central repository rather than keeping back seed from year to year, and the technical advancements in the development of new strains and varieties for adoption by farmers that have stemmed from the Green Revolution, have facilitated the very rapid adoption and dissemination of GMOs globally. This will narrow the range of available genetic variation even further, as even fewer individuals (often just one) will likely be used to define and make up all if not most of the following generations. In providing farmers an economic incentive to adopt GM foods by making them cheaper and less labor intensive to produce than their non-GM counterparts, and in increasing the incentive to consumers of values added (added nutrients, less herbicide and pesticide contamination), GMOs are likely to further constrain an already exaggerated genetic bottleneck by further limiting the genetic variation that goes to establish the following generation.

GM Foods 2: Which Types of Genes and Why

The current GM foods undergoing testing and regulatory approval are usually those considered to confer an advantage under present day agricultural management practices. Examples of such GM foods therefore span a range from those with traits that can allow a plant to utilize less of an existing resource to those that allow a plant to function better in a particular environment. A third category includes foods with traits that are seen as value added (e.g., providing an additional needed nutrient). Sources such as U.S. Field Test Applications[13] or Canadian Novel Food Approvals[14] can provide specific examples of current GM food crops.

In the case of herbicide-resistant crops, the notion to limit the use of the herbicide and yet also more efficiently kill weeds that would otherwise compete with the cash crop for nutrients is desirable from a farmer's vantage. The attraction of a herbicide-resistant crop is that a single application of herbicide while the crop is in the field replaces the alternative of first applying a herbicide to kill weeds, then planting a crop, and then reapplying herbicide after the harvest. Two or even three applications of herbicide are said to be replaced with one, and the result is less overall use of herbicide. This would translate into less cost in supplies and labor to the farmer without compromise in the quality of the crop produced and would also lessen the environmental impact posed by repeated applications of herbicide. This type of strategy was undertaken in achieving resistance to predation by insects through the expression of a pesticidal bacterial gene, Bt, which now has been introduced into a variety of crops.

There is evidence that these strategies have been economically successful, at least in that farmers have adopted them at increasing rates in the past several years, resulting in higher yields, less use of pesticides, and higher profits.[15] Ecological concerns such as negative effects on nontargeted insect species seen in laboratory simulations have not yet been demonstrated in real-life field conditions,[16] although there may be some evidence that spread of Bt toxin to nearby nontransgenic crop insect refuge areas may be undermining attempts to limit development of Bt-resistant insect strains.[17] Unfortunately, much of the literature still reflects the biases either of scientists committed to the success of their creations or environmentalists opposed to their adoption under any circumstances, and a clear picture determined by independent observers has yet to emerge.

Producing crops that have added nutritional benefit to the consumer (and thus a potential market) is being pursued as a way to address nutrient deficiencies in diets around the world. The crops chosen are typically those normally grown in the region where the particular nutritional benefit is seen to be deficient. The assumption is that this will allow more readily the adoption of the GM food both culturally and from a standard-of-practice viewpoint. Examples of this are GM rice varieties that have been engineered to increase the amount of iron to combat anemia, or vitamin A to eliminate this deficiency as the root cause of blindness in certain regions of the world. Other approaches, such as GM foods that allow for greater protein content or drought resistance, have the potential to provide locally grown products of high nutritional value in parts of the world where they are desperately needed and would otherwise be difficult to produce. Whether the result will live up to the promise is yet to be seen.

GM Foods 3: The Future

There are many simple traits that still have the potential to be exploited in a GMO context. With DNA technologies able to facilitate identification

and isolation of genes from any organism at a phenomenal rate, the world
has now become open to any and all prospects of biomining (seeking unique
genes for traits of commercial/social importance). For the immediate future,
the biotechnology industry will continue to seek new ways to exploit existing agricultural practice while potentially adding value to the foods we consume for both the farmer and the consumer. Additionally, newly developed
GM foods are likely to undergo a refinement over existing technical strategies with respect to the specific tissue targeted for transgene expression as
well as the permitting of only transient expression of the transgene. Much
of this will be mediated through the regulatory domains of genes being
expressed. If the specific part of the food being consumed never expressed
the transgene in question this might help allay some public concerns with
respect to consuming one food containing a protein derived from a different
organism. Other issues would still remain, however, as the GMO in question
would still contain the DNA of a transgene in every cell.

Other innovations in GMOs are likely to involve addressing more
complex traits. To date this arena has been the purview of traditional breeding programs that only permit those with the desired complex trait to give
rise to the next generation. The shortfall of this method is that it involves
the reshuffling of two whole genomes, nature's way of generating variety
by "shuffling the deck" as genes are passed on from one generation to the
next. Referred to as the "independent assortment of genes," this trick of
nature is meant to result in generating as much variation as possible for the
prospect of producing the best candidate for selection into future generations in response to the selection pressures that might prevail. Keeping a
trait intact is by definition, therefore, not guaranteed, as the "independent
assortment" of the alleles that contributed to the desired trait in the one
individual will likely segregate over time and across successive generations
necessitating a constant monitoring in breeding programs (a selection pressure often called "artificial selection"). This issue is typically of less concern
with plants where self-fertilization is often at play in helping a complex trait
breed true. In animals, however, a constant monitoring of herds is required
to maintain the traits desired.

As more is understood about complex traits, the ability to manipulate them (and by definition multiple genes simultaneously) will likely also
become a target for the development of GM foods. One potential driver
for this innovation will likely be new technologies such as artificial chromosomes that can contain and simultaneously express all the genes contributing to a complex trait. Some part of this will be driven by the need
to secure niche markets involving patenting of GMOs difficult to reproduce
by competitors and by traditional methods as described earlier. The future

of such pursuits likely will entail genetic modification techniques that will result in a significantly larger percentage of the genome owing its makeup to many transgenes.

GM Foods 4: What Are the Real Risks?

As former U.S. Secretary of Defense Donald Rumsfeld famously observed at a Pentagon press briefing: "There are known knowns. There are things we know that we know. There are known unknowns. That is to say, there are things we know that we don't know. But there are also unknown unknowns. These are things that we don't know we don't know."

There is much debate as to the potential for GM foods in creating greater efficiencies, lesser toxicity, and greater yields and providing much needed nutrition to parts of the world where cultural and economic factors work against the indigenous populations. Many of these debates will rage on regardless of the development of the science. In genetic terms, much of the detriment sustained by loss of genetic variation or biodiversity has already come to pass through current agricultural practice. In generating GM foods it is argued that we run the further risk of randomly altering existing genes in an organism by the disruption or mutating of genes through techniques that yield a higher spontaneous mutation rate (gene transfer, somatic cell selection, radiation, chemical mutagenesis) than is found in nature.

The Problem of Diversity or the Lack Thereof

Diversity or the lack of uniformity is judged by the standards of modern day agricultural practice to be grossly inefficient. Uniformity in performance is a hallmark of the many plants and animals produced under the stringent conditions of present day agricultural practice. GMOs, as noted previously, herald a further exaggeration of this "trait." Will a further narrowing of the genetic bottleneck through selection via yet fewer individuals be the genetically modified straw that breaks the proverbial transgenic camel's back? Modern agriculture, whether it be animal husbandry or the growing of crops, has increasingly come to rely on rigorous management practices that have been widely adopted since they result in consistency of product quality and yield. However, their reliance on only a few individuals or varieties that are used to give rise to all successive generations brings with it a confounding and worrisome issue in the long run. The promise of GM foods in offering yet further advantages to farmers and consumers is poised to result in an even greater exaggeration of the problem of fewer varieties and even lesser genetic variation.

Defined by "Our" Genes

Does the presence of a gene/protein originally derived from another organism (even if the protein expression is absent in the tissue consumed) change an organism sufficiently to label it "abhorrent"? One response to this concern has been to point to the fact that "foreign" genes are often introduced into the genomes of many organisms by viruses and bacteria as a consequence of an infective cycle. A fallacy to this argument is the scale to which this occurs (i.e., not across all individuals in a population) and the "sameness" (both in the genetic material introduced and in the sites in the host genome that are perturbed when such "foreign" material is introduced) of its affect from one individual to another in a given species (i.e., heterogeneity or variation is still maintained).

It is also necessary to have some understanding of the composition of different genomes. In contemplating the relationship between humans and chimpanzees, it has been estimated that there is an approximately 1.5 percent difference between the two species in all of the proteins encoded by our respective genes.[18] The amount of structural or coding sequence in our genome (human) for the approximately thirty thousand proteins we are made up from is estimated at 1.2 percent of all the genetic material we inherit. The remaining 98.8 percent contains the regulatory elements and many other features in DNA that affect the ordering and organization of our genetic material and its regulation. Therefore, it is largely this regulation, ordering, and organization of our genetic material that would seem to confer on us the properties of being human and not chimpanzees.[19] The result of the organization wrought by our genome versus that of a chimpanzee is the differential expression of many thousands of genes in time and space. The cumulative effects of this would seem to conspire to produce the final result of one organism as distinct from the other. It is also the similarity in some of the organization of the relative genome of humans and chimps that confers the degree of similarity we share as opposed to humans and sheep. Thus if it is the organization of our genomes that to a large extent determines what we are, does the alteration of this organization by the introduction or change to a single gene constitute a significant change? What effect does such a gene, via its regulatory or structural domains, have on the nature of development in an organism? Are we not able as yet to measure changes that are too subtle to observe by our instrumentation or to pose stresses that will bring out differences not contemplated? How many such changes does it take to consider an organism no longer what it was before we began our manipulations?

Summary

- Genes are more complicated than most people commonly think, involving as they do complex regulatory as well as structural information.
- This complexity makes it difficult to accurately predict the impact of genetic alterations. This is particularly the case in nontransgene GMOs where desired changes are created using techniques that cause random mutations to the genome.
- At the level of the chemical constituents of DNA few distinctions can be made among plants, animals, and microbes. The consequence is the relative ease with which genes from one organism can be expressed in another provided the appropriate regulatory sequences are in place for expression in the GMO.
- Genetically determined characteristics may be the result of few genes (simple traits) or many genes (complex traits). Commercially important simple traits have been the focus of GMOs to date.
- Transgene GMOs have arisen from a utilitarian view of simple traits and genes conferring the trait.
- A lot of scientific energy has been expended determining how/which genes are related to which traits; in learning to manipulate genes, we have learned how to manipulate traits in a new way. The future of transgene GMOs will draw from this knowledge, as compendiums of simple traits or else all alleles contributing to a complex trait can be successfully placed in one organism using new molecular genetic techniques (e.g., artificial chromosomes).
- Human beings have introduced considerable genetic modification into agricultural species by selective breeding (the old way of manipulating traits) for desired traits. This represents thousands of years of genetic manipulation further amplified by modern day agricultural practice that, while broadly seen as acceptable, has become the springboard for the rapid acceptance and dissemination of GMOs.
- Adoption of these large-volume, uniform production methods has led to a decrease in the genetic diversity of agricultural products. A significant feature of the practice of modern day agriculture, the continual desire for better, cheaper food, is driving the pressure toward the continued

development of new agricultural technologies. These new ways of manipulating traits (GM foods) will likely further decrease this genetic diversity.

Notes

1. Lander et al., "Initial Sequencing and Analysis of the Human Genome," *Nature* 409, no. 6822 (February 15, 2001), pp. 827, 860–921; Venter et al., "The Sequence of the Human Genome," *Science* 291, no. 5507 (February 16, 2001), pp. 1304–51.

2. R. Mayeux, "Mapping the New Frontier: Complex Genetic Disorders," *Journal of Clinical Investigation* 115, no. 6 (June 1, 2005), pp. 1404–07; C. Turnbull, "Genetic Predisposition to Cancer," *Clinical Medicine* 5 (2005), pp. 491–98; Online Mendelian Inheritance in Man, OMIM (TM). McKusick-Nathans Institute for Genetic Medicine, Johns Hopkins University (Baltimore, MD) and National Center for Biotechnology Information, National Library of Medicine (Bethesda, MD, 2000). http://www.ncbi.nlm.nih.gov/omim/.

3. R. J. Coakley, et al., "Alpha1-antitrypsin Deficiency: Biological Answers to Clinical Questions," *American Journal of Medical Science* 321, no. 1 (January 2001), pp. 33–41. http://www.alphaone.org.

4. E. R. Ahrens, V. V. Gossain and D. R. Rovner, "Human Insulin: Its Development and Clinical Use," *Postgraduate Medicine* 80, no. 1 (July 1986), pp. 181–84, 187.

5. Food and Agriculture Organization of the United Nations, "Genetically Modified Organisms, Food Safety and the Environment" (Rome, 2001). ftp://ftp.fao.org/docrep/fao/003/x9602e/x9602e00.pdf.

6. Health Canada, Guidelines for the Safety Assessment of Novel Foods. http://www.hc-sc.gc.ca/food-aliment/mh-dm/ofb-bba/nfi-ani/e_nvvliie01.html.

7. Health Canada, Health Products and Food Branch, Novel Food Decisions. http://www.hc-sc.gc.ca/food-aliment/mh-dm/ofb-bba/nfi-ani/e_nf_dec.html.

8. Jared Diamond, *Guns, Germs and Steel: The Fates of Human Societies* (New York, NY: Norton, 1997, 1999), pp. 83–191.

9. Mark Winston, *Travels in the Genetically Modified Zone* (Cambridge, MA: Harvard University Press, 2002), p. 20.

10. G. S. Khush, "Green Revolution: Preparing for the Twenty-First Century," *Genome* 42 (1999), pp. 646–55.

11. Mark Winston, *Travels in the Genetically Modified Zone* (Cambridge, MA: Harvard University Press, 2002), p. 25.

12. Clive James, "Preview: Global Status of Commercialized Biotech/GM Crops: 2004," *ISAAA Briefs No. 32, The International Service for the Acquisition of Agri-biotech Applications*, Ithaca, New York.

13. U.S. Department of Agriculture, National Biological Impact Assessment Program, ISB database *"Field Test Releases in the US."* http://www.nbiap.vt.edu/cfdocs/fieldtests1.cfm.

14. Health Canada, Novel Food Decisions.

15. See G. Traxler, "The GMO Experience in North and South America: Where to from Here?" Proceedings of the Fourth International Crop Science Congress (September 26–October 1, 2004), Brisbane, Australia. www.cropscience.org. au; M. Scott and L. M. Pollak, "Transgenic Maize," *Starch* 57, no. 2 (May 2, 2005), pp. 187–95.

16. Y. Shirai, "Influence of Transgenic Insecticidal Crops on Non-Target Arthropods: A Review," *Japanese Journal of Ecology [Jap. J. Ecol.]* 54, no. 1 (2004), pp. 47–65.

17. C. F. Chilcutt and B. E. Tabashnik, "Contamination of Refuges by *Bacillus thuringiensis* Toxin Genes from Transgenic Maize," *PNAS* 101, no. 20 (May 18, 2004), pp. 7526–29.

18. H. Watanabe, et al., "DNA Sequence and Comparative Analysis of Chimpanzee Chromosome 22," *Nature* 429, no. 6990 (May 27, 2004), pp. 382–88; The Chimpanzee Sequencing and Analysis Consortium, "Initial Sequence of the Chimpanzee Genome and Comparison with the Human Genome," *Nature* 437 (September 1, 2005), pp. 69–87.

19. M. C. King and A. C. Wilson, "Evolution at Two Levels in Humans and Chimpanzees," *Science* 188, no. 4148 (April 11, 1975), pp. 107–16.

2

Ethical Perspectives on
Food Biotechnology

Paul B. Thompson

The use of recombinant DNA to modify the genetic structure of plants, animals, and microbes (see chapter 1, by Abraham) and the ability to clone adult cells from mammals jointly contributed to an international controversy that has several axes of contention. While theologians and philosophers have thus far focused primarily on applications in the field of human medical science, the broader public has arguably been equally concerned with the use of these techniques in food and agriculture.[1] This popular concern with biotechnology (as the techniques of gene modification and adult cell cloning will henceforth be called) reflects a worry that the technology may have unknown and unacceptable risks, but also apprehension about ethics.[2]

On the one hand, ethics deals with almost universally recognized norms that are both implicit within everyday social interaction and explicitly articulated in public sources such as legal or professional codes of practice, religious texts, folktales, literature, and philosophy. On the other hand, the ethical dimensions of conduct and reflection are often characterized as inherently personal, introspective, and inherently unsuited to public discourse. Ethical concerns associated with food and agricultural biotechnology can thus be expected to comprise highly idiosyncratic personal reactions of individuals, identifiable traditions and values of particular social groups, and broadly shared social norms. The goal in this chapter is to summarize and articulate rationales behind using, opposing, or qualifying the use of agricultural biotechnology as they have been expressed in nearly twenty years of debate.

One way to approach this task is to present the debate in terms of opposing "pro" and "con" arguments, as several studies by philosophers have done. Gregory Pence, for example, emphasizes the way proponents of biotechnology emphasize humanitarian goals such as ending hunger, while opponents see biotechnology as unnatural, a "mutant harvest."[3] Pence's focus on naturalness was also the main organizing principle for an earlier study by Michael Reiss and Roger Straughan that included medical as well as agricultural biotechnology.[4] Gary Comstock also takes up the possibility that biotechnology might be unnatural but emphasizes how he himself came to see the humanitarian rationale for biotechnology as overriding his own concerns about the social and environmental risks associated with transgenic crops and genetically engineered animal drugs.[5] Interestingly, all these authors wind up on the "pro" side of the debate. This way of framing the debate in terms of benefit from increasing agricultural productivity, on the one hand, and unnatural or risky technology, on the other, has also been the subject of a lengthy and careful study by Hugh Lacey, who is less inclined toward the "pro" point of view. Lacey believes that the pro-biotech perspective is rooted in an ethical perspective that valorizes processes of control and predictability, while the anti-biotech perspective can be traced to scepticism about the viability and desirability of control.[6]

This chapter interprets controversy over agricultural biotechnology as an episode in several ongoing and overlapping social, political, and ethical struggles over guidance of food and food production. Because food consumption is both rich in symbolic or cultural significance and biologically necessary for human life, any technology for producing or preparing food has ethical ramifications of one kind or another. These include the way that the technology affects safety and access to food, as well as other questions of fairness and equity associated with the broad system for producing and distributing food. As will become clear in this chapter, much of the debate involves ethical matters that could be raised for any food technology. They are issues of *general technological ethics*. Yet there *are* characteristics of biotechnology that create forms of ethical apprehension that do not arise in connection with chemical, mechanical, and other food technologies. These can be referred to simply as "special concerns." Some of them indeed overlap with questions in biomedical applications of genetic technology.

General Technological Ethics

The twentieth century was a time of unsurpassed technological progress, but it was also a time in which humanity learned that technological changes bring unintended social and environmental consequences. The German philosopher Hans Jonas is generally credited with first recognizing the need for an ethical approach to anticipating and evaluating technology. He argued that technological ethics must integrate science-based attempts to understand the systematic and temporally distant effects of technology with ethical concepts attuned to the fact that many of the people who will be affected by technology will not be known to those who plan and execute a technological practice. Jonas called for a "principle of responsibility" (*Prinzip verantwortung*) as a response to this situation.[7]

Jonas's approach does not point us toward the pro and con weighing of biotechnology undertaken by Pence, Comstock, or Lacey. The *Prinzip verantwortung* calls for scientists and engineers to make an active attempt to anticipate possible forms of harm. But if they do this and find no harm, they are at liberty to proceed with their technology. The expectation that technological innovations will improve workplace efficiencies (lowering consumer cost) provides a broad argument that further supports them doing so. There is, in short, no real need for a "pro" argument for biotechnology or any other technology, at least not at the outset. There is only the need for a responsible effort to ascertain the unintended consequences of technical change. The debate on agricultural biotechnology revolves around five general categories in which the products and processes of rDNA have been alleged to pose risk: (1) impact on human health (i.e., food safety); (2) impact on the environment; (3) impact on nonhuman animals; (4) impact on farming communities in the developed and developing world; and (5) shifting power relations (e.g., the rising importance of commercial interests and multinationals).

Food Safety

Critics of food and agricultural biotechnology often link the need for ethics with a concern for food safety. This is, on the one hand, quite understandable, since if one already believes that eating so-called GMOs (genetically modified organisms or the products of food and agricultural biotechnology) could be dangerous, one is also very likely to believe that it is unethical to put people in a position where they might eat them, especially without their knowledge. On the other hand, those who advocate on behalf of agricultural biotechnology take offence at this characterization of ethics, since it implies that they are exposing the unwitting public to grave dangers without their knowledge. In fact, what is at issue between critics and advocates of biotechnology is often not really a question of ethics. Both would agree that it would be very unethical to expose people to food-borne hazards without their knowledge. The source of their disagreement is whether there *are* hazards associated with the human consumption of GMOs or, if harms are theoretically possible, the likelihood that any potential hazards will actually manifest themselves in the form of an injury to human health.

One ethical issue concerns the question of what a company or government food safety regulator should do when there are disagreements of the sort just mentioned. One possible answer is that the decision should be based on the best available science. If GMOs have benefits of some sort (if only the potential to increase the cost efficiency of crop production and build wealth for farmers and seed companies), then the ethical rationale for this approach presumes that it would be wrong to prohibit GMOs without some sort of evidence that they pose hazards to human health or to the environment. If one allowed baseless concerns to stifle innovation, the result would be technological and economic stultification that is not in the public interest. This approach does require criteria for deciding when an alleged hazard is baseless. "The best available science" is then supposed to provide a risk-based approach (discussed later) to this problem.[8]

Even under the best circumstances of strong scientific consensus on hazards, this approach to food safety suffers from some of the problems often associated with the utilitarian or consequentialist form of ethical reasoning with which it is closely allied. Any approach to ethics that rationalizes some chance of a hazardous outcome in terms of benefit to the general public will be vulnerable to criticisms that stress individual rights.[9] The risk of allergic reaction to GMOs is an instance of this problem. Since genes make proteins, and proteins are potential allergens, one cannot exclude the possibility that genetic engineering of foods may introduce proteins into foods that will cause sensitivities and allergic reactions in some portion of the population. Since food allergies are not well understood, and since they may affect very small percentages of the population, it may not be practical to anticipate or characterize the likelihood of allergic reactions before GMOs are released for public consumption. Thus, there may be a few people who would be harmed by eating a GMO, and the mainstream approach to food safety described above seems to rationalize a small probability of serious health affects on these few in terms of economic benefits to the many.

In contrast to this view, one may think that individuals have a right not to be harmed by inadvertently consuming a protein that they could not have known they were allergic to or that this right is violated even when the risk is uncertain. But do people who are not concerned about speculative or low-probability risks have a right to enjoy the benefits of new technology? One way to protect both kinds of right is to place each individual in a position to look after his or her own interests where food safety is concerned. This approach follows the ethical logic of informed consent: people should be free to take whatever risks they choose, but they should not be put in a position of risk without adequate notification and an opportunity to choose otherwise. This sort of reasoning has led many to demand labels for GMOs.[10] However, the informed consent approach to food safety has drawbacks. Gary Comstock discusses empirical research showing how apparently detailed food information can distort personal decision making. It may be impossible to provide the information that allows one person to make an informed choice without simultaneously putting another person in a position where he or she will make an uninformed choice. As such, one might argue that governments should be judicious and sparing in the information that they require to be supplied to consumers, and this argument effectively brings us back to the "best scientific evidence" perspective described already.[11] Thus, food safety *is* embroiled in ethical debates, despite broad agreement that it would be unethical to knowingly place people in a position of consuming unsafe food.

Ethical Significance of the Environment

While there are many ways in which environmental responsibilities might be interpreted, one central and abiding ethical question unifies a host of approaches with the hazard-identification phase of mainstream risk assessment: what counts as an ethically significant environmental impact? One useful way to summarize issues arising in connection with environmental impact is to note that answers to this question can raise three different kinds of ethical concern. First are human health effects accruing from environmental exposure, such as air- or water-borne pathogens (as opposed to ingestion through food). Second are catastrophic impacts that would disrupt ecosystem processes in ways that threaten to destabilize human society. This includes dwindling energy supplies, human population growth, and global warming. Finally there are effects that are felt less by humans than by the broader environment, that is, ecocentric (or nonanthropocentric) impacts. Interpreting each of these three types of environmental impact as having ethical significance involves distinct ethical concepts and values, some widely endorsed and others less so.

Environmental impacts in the first category manifest themselves as human injury or disease. They include cancer induced by chemical pollution, emphysema and lung diseases from air pollution, poisonings, and nonfatal diseases such as allergies and reduced fertility speculatively associated with hormone-disrupting chemicals in the environment. Although the scientific and legal issues that arise in establishing the connection between cause and effect are tortuous, the ethical imperative to limit these risks is very clear. Ethical and quasiethical issues arise because it is not clear how to resolve uncertainties that arise in assigning a probability to the unwanted impact and because there are different ways to think about the social acceptability of environmental exposure to human health risks. Although it is certainly possible that food and agricultural biotechnology could pose such risks, products currently under development for use as food have not been linked to any known human diseases that would be contracted by environmental exposure. Critics of biotechnology have noted that transgenic crops are also being developed to produce drugs and industrial products and that these products must be contained in order to limit environmental exposure to human health hazards.[12] No one has contested the claim that hazards to human health through environmental exposure are ethically undesirable.

For many years, the environmental risks associated with agricultural biotechnology were thought to fall primarily in the middle category of potentially catastrophic ecological consequences. In contrast to environmental exposures that might lead to human health hazards, the science that would be used to predict and measure the likelihood of ecological catastrophe is less well developed. Ecologists raised the possibility of widespread disruption of atmospheric processes associated with ice-nucleating bacteria early in the development of agricultural biotechnology.[13] The speculation that biotechnology would contribute to a narrowing of the genetic diversity in major food crops was also an early concern.[14] During the 1990s the potential environmental impacts foreseen were less sweeping. Particular attention has been given to the potential for escape of herbicide-tolerant genes into weedy relatives of crop plants and to the possibility that insect pests will acquire resistance to *bacillus thuringiensis* (bt).[15] Though such events are not in themselves catastrophic, their ethical significance derives from interpreting them as contributing to a broad destabilization of the global food system.

North American philosophers writing on environmental ethics have laid greater stress on duties to nature than on duties to posterity, suggesting that the third category, of ecocentric or nonanthropocentric effects, might be of particular ethical significance. Although questions of uncertainty and risk acceptability might also arise in connection with impacts on wildlife or ecosystems, here there is more debate over why such impacts might be thought to have ethical significance. Preservation of wilderness and endangered species has been of particular importance in Canada and the United States. In part, this emphasis derives from the fact that environmentalists in Canada and the United States have sought persuasive rationales for setting aside the relatively large tracts of undeveloped land that exist in these countries. Industrial, scenic, and recreational uses provide a baseline for valuing wild ecosystems in economic terms. The main philosophical tasks have been understood in terms of developing a rationale for valuing and preserving wild ecosystems, including keystone species, irrespective of their economic value. Given this orientation, one would expect that products such as transgenic salmon, which could affect wild salmon populations, would be among the most contentious applications of biotechnology from the perspective of ecocentric environmental ethics.

In addition, agriculture is sometimes viewed as antithetical to environmental values in the North American context. Agricultural technologies are potential polluters, contributing to human health risks, and agricultural land use competes with wilderness preservation. For example, Canadian environmental ethicist Laura Westra argues that farmlands cannot possess "ecological integrity." She sees farming as environmentally valuable only as a buffer that protects wild areas from the impact of human civilization.[16] Given this orientation, one might think that agricultural biotechnology would not be of interest on ecocentric environmental grounds. A contrasting view, which may be more prevalent in northern Europe, implicitly sees preservation of nature as preservation of farmland. Preservationist goals are articulated in terms of keeping land in fairly traditional forms of farming, and farming is seen as wholly compatible with preservation of habitat.[17]

Prior to 1999, crop biotechnology was not widely associated with environmental impacts on wilderness or endangered species. In that year news reports that Bt crops could affect monarch butterflies enlivened the prospect of unintended impact on nontarget species for the first time. This has awakened public recognition of the way that agricultural biotechnology could have an impact on wild species and provides an example of how ecocentric environmental impacts could be brought about by genetic agricultural technologies. In Canada, genetically engineered canola could outcross with wild rape. Research on genetically engineered fish has long been associated with the potential for negative impact on wild populations. There are also less well known products, such as recombinant vaccines, that could have a negative impact on wild habitats as well. As experience and experimental studies accumulate, the list of possible hazards is expanding, and scientists' ability to quantify the likelihood that such hazards will materialize is increasing.[18]

An additional type of environmental impact requires one to see a farmer's field as having a kind of ecological standing or integrity of its own. Critics may see biotechnology as threatening in virtue of the possibility that transgenic plants may appear in a field in which a nontransgenic crop is growing, either by pollen drift, contamination of the seed supply, or when volunteer transgenics survive over the winter to reappear in a field sown to nontransgenics in the succeeding year.[19] The key philosophical question is, why does this matter? Some answers to this question are economic. A farmer may lose the ability to gain a price premium for a nontransgenic crop or in the worst case lose the ability to sell the crop in some international markets altogether. Here, an ecological or environmental mechanism contributes to an impact that is better classified as "socioeconomic" than "environmental."

Other answers draw upon the European tendency to see farms as part of nature. Still other answers may foreshadow the discussion of purity and unnaturalness that is taken up in the section on special concerns.

Moral Status of Animals

Like impact on ecosystems or ecosystem processes and unlike impact on human health, the impact of human action on nonhuman animals is controversial because some people deny that animals can be harmed at all. The belief that animals are nonsentient machines who feel no pain is often attributed to René Descartes (1596–1650) and has without question been influential in the use of animal experimentation within the medical sciences.[20] Immanuel Kant (1724–1804) believed that animals could not be harmed because they lacked reason and argued that the moral wrong associated with animal abuse owed not to any harm suffered by the animal but solely to the harm that a perpetrator inflicts upon himself in acquiring a habit of poor character.[21] Long before these pivotal figures in European philosophy, philosophers of the ancient world such as Aristotle had defended the view that animals lack the mental faculties that would make human conduct toward them morally significant,[22] and Thomas Aquinas had written that if the Bible appears to forbid cruelty to animals it is only to guard against the possibility that "through being cruel to animals, one becomes cruel to human beings; or because injury to an animal leads to the temporal hurt of man."[23]

The philosophers Peter Singer and Tom Regan have jointly campaigned against the belief that animals do not count morally, arguing that these philosophical attitudes are causally responsible for untold amounts of animal suffering in medical research, product testing, and animal agriculture. Singer and Regan oppose one another, however, in offering accounts of why animal suffering is morally significant. Singer places his argument within the tradition of utilitarian philosophy, often quoting the venerable founder of this tradition, Jeremy Bentham (1748–1832), who wrote, "[T]he question is not, Can they *reason*? nor, Can they *talk*? but, Can they *suffer*?"[24] Here, the ethical basis for concern about the impact of human activity on nonhuman animals follows from the utilitarian mandate to act in ways that maximize the ratio between pleasure or satisfaction and pain or suffering. If animals experience pain (and Singer produces prodigious amounts of scientific data and argumentation to support the commonsense belief that they do), then we are morally obligated to take their pain into account when evaluating our actions in ethical terms.[25]

Regan presents his argument in favor of animal rights by arguing first against the general utilitarian framework that Singer accepts. Instead, Regan believes that the only philosophically defensible approach in moral philosophy is to observe rights that protect the interests of moral subjects. In Regan's view, the key philosophical burden of proof involves whether or not animals possess the interests characteristic of moral subjectivity. These involve having a sense of oneself, a continuous form of conscious experience capable of supporting a minimal experience of personal identity. Regan defends the view that vertebrate animals, at least, do indeed possess such interests and are, in his terminology, "subjects-of-a-life" or bearers of a cognitive identity commanding our moral respect.[26] Sociologists James Jasper and Dorothy Nelkin give Singer, Regan, and other philosophers great credit for initiating the worldwide social movement for reform in a number of domains in which animals are used by human beings.[27]

The philosophical movement for animal welfare (Singer's view) and animal rights (Regan's view) has broad implications for agriculture and for biotechnology. Both Singer and Regan advocate vegetarianism, for example, though Singer's commitment to vegetarianism might wane were it possible to institutionalize more humane forms of animal production. However, an even cursory discussion of animal ethics as they apply within contemporary animal production settings would take the present discussion far afield, and many of these topics will be covered in chapter 3 by Lyne Lètourneau. It must suffice to note that public interest groups advocating humane treatment of animals monitor developments in animal agriculture closely. They also take a keen interest in biotechnology, though here much of the debate has focused on transgenic mice developed for biomedical research.[28] The balance of the discussion in this chapter will narrow these broad concerns to the topic of biotechnology applied to food animals.

Genetic transformation and cloning of livestock are currently in the experimental stage. Nevertheless, survey research indicates that animal biotechnology is strongly associated with ethical concern among members of the public.[29] There are also a number of authors associated with social movements to protect animals who have decried food and agricultural biotechnology.[30] However, other authors who have argued strongly for recognition of animal interests have not found gene technology to be especially problematic.[31] Clearly some of those who find animal genetic engineering problematic are among those who see gene technology as intrinsically wrong, a topic treated as a special concern discussed below, but gene technology applied to animals raises two additional issues that might also be applied to animal breeding and that thus belong in the category of general technological ethics. The first is that gene technologies have the potential to produce suffering in animals. The second is whether or not it is acceptable to reduce an animal's capacity to suffer as a means to reduce suffering.

Some of the first genetically engineered animals were very dysfunctional,[32] and there continue to be questions about the health of cloned animals (though the evidence currently suggests that they do not have abnormal health problems). Animals have not always and everywhere been thought to have moral standing that would make their suffering a matter of ethical concern. Nevertheless, few in Western industrial democracies would deny that animals are capable of feeling pain, and few would deny that humans have a responsibility to ensure that animals do not suffer gratuitously. The ethical issue here is thus whether the purposes to which animals are being put justifies any pain and suffering they experience. Although this is an ethical issue of general interest and importance, its bearing on the ethical acceptability of animal biotechnology should not be overstated. No genetic transformation that would result in genetically engineered or cloned animals enduring greater suffering than ordinary livestock is being proposed. Bernard Rollin has argued for an ethical principle that would proscribe any such application of biotechnology. To the extent that existing practices within livestock production are ethically acceptable with respect to their impact on farm animals, practices associated with food and agricultural biotechnology should also be acceptable.[33]

Of course, existing practices are the subject of intense criticism by animal advocates, and arguments that follow the principle stated in the preceding paragraph have already been controversial. For example, recombinant bovine somatotropin (rBST), a product of genetically engineered bacteria that stimulates dairy production, has been controversial because cows with higher rates of milk production are also at a higher risk for health problems. The U.S. Food and Drug Administration chose to interpret the animal health risk from use of rBST as consistent with that of existing practices, since there are other legal ways for boosting milk production. Critics chose to interpret the same data as evidence that rBST increases the risk of health problems in animals on which it is used.[34] There is thus a real prospect that animal advocates will interpret the animal health risks associated with gene technology as having greater ethical significance than that of existing technology.

The second set of ethical issues associated with animal biotechnology was first clearly stated when Rollin suggested that genetic engineering should be used to render animals being used in medical experiments "decerebrate"— physically incapable of experiencing pain.[35] This general approach could be applied in a less drastic fashion to livestock. Gene technology could be used to produce animals that are more tolerant of the crowding and confinement that create welfare problems in existing animal production systems. It is, in fact, possible to do this through conventional animal breeding, so this consequence should not be seen as uniquely associated with recombinant gene transformations.[36] If animal suffering is the predominant ethical concern, it would seem that there is a compelling ethical argument for doing this. Many animal advocates find this to be an abhorrent suggestion, though it has proved difficult to articulate reasons that do not revert back to the claim that animals have a form of telos, or intended design. This notion of telos has been cited by a number of critics who find genetic engineering of animals to be intrinsically wrong. These arguments are discussed below as a form of concern special to biotechnology

Impacts on Farms and Farm Communities

Agricultural production technology affects economies of scale in farming and food distribution and also affects the control that different persons or groups maintain with respect to the overall food system. Certainly any technology has these effects, including not only such obviously agricultural technologies as plant breeding or chemical pesticides but also information technologies such as the internet and basic infrastructure such as roads and transport. How do technological changes pose challenges of social justice with respect to farming communities? Perhaps more than any of the other ethical concerns discussed in this chapter, food and agricultural biotechnology represent nothing more than a case study for this general question.

The debates over transformation of rural areas in Europe between 1600 and 1800 were generalized and evolved into comprehensive views on social justice during the nineteenth and twentieth centuries. Arguments that favored agricultural technology eventually took shape as the neoliberal principles endorsing the social efficiency of unregulated markets, on the one hand, and the sanctity of private property, on the other. Arguments opposing technological improvement of agricultural production and rural infrastructure evolved into socialist and communitarian conceptions of social justice aimed at preserving (or reinstituting) social institutions that protected the poor. The antitechnology dimension of these arguments was gradually muted, particularly in strong leftist and Marxist interpretations of social justice. There is thus a sense in which some of the broadest concepts of social justice have their roots in disputes over agricultural technology. Disputes over agriculture and rural development continued throughout the twentieth century, but participants in these debates were not particularly mindful of their historical origins. It is useful to isolate two themes.

First, new agricultural technology had its greatest effect on rural communities in industrial societies during the twentieth century and especially after World War II. This created a century-long debate over the ethical and political wisdom of allowing industrial principles to shape agricultural production versus policies and technological investments that would strengthen family ownership structures and rural communities. The ethical dimension of the debate consists in the claim on the one side that technological innovations adopted by profit-seeking farmers, processors, and food retailers reduce overall food costs, resulting in consumer benefits that outweigh the financial and psychological costs of those who suffer economic reverses. On the other side it is claimed that the economic opportunity represented by family farms and the small businesses that arise to support them is the essential component of social justice. Furthermore it is claimed that small-scale rural communities promote participatory local governance and are therefore most consistent with the ethical principle that social justice depends upon consent of the governed.[37] It was virtually inevitable that any new agricultural technology developed in the last quarter of the twentieth century would be subsumed by this debate. Some of the first social science publications on food and agricultural biotechnology framed it in precisely the terms of the century-long debate over the structure of agriculture and the ethical importance of the family farm.[38]

A second strand of ethical concern over social justice examined the impact of food and agricultural biotechnology in developing countries. Here, too, there was an ongoing debate over the "Green Revolution" agricultural development policies being pursued by organizations such as the World Bank, FAO, the Consultative Group on International Agricultural Research, the Rockefeller Foundation, and the international development agencies of industrialized nations. Like the first strand of debate, critics of the Green Revolution have argued that increases in agricultural productivity have been gained at the expense of rural ways of life, a repeat of failures and tragedies that have faded from the memory of people in the industrialized world. Here, too, it was inevitable that biotechnology would be subsumed by the existing debate.[39]

On the part of those who support the actions of the official development organizations, it is argued that developing countries must follow the lead of the developed world in adopting yield-enhancing agricultural technology. As earlier, it is argued that the benefits of increased food production outweigh any short-run reverses suffered by individual farmers. Indeed, given the threat of famine, it is argued that the social demand for more food production is compelling.[40] Those holding an opposing view raise factual questions about the success of the Green Revolution. The ethical dimension of their viewpoint notes that the infusion of technology and capital into peasant economies and traditional agricultural production systems causes an upheaval in the existing social relations. In addition to claiming that this upheaval destroys the culture and way of life in traditional societies, critics of Green Revolution-style development note that the poorest of the poor are the most vulnerable when such massive transformations of social structure occur. They counter the argument that food needs in the developing world override concern for cultural integrity with an argument that appeals to the basic rights of individuals whose lands, jobs, and way of life are destroyed in the wake of development projects.[41] These general criticisms have been extended to biotechnology in a series of critical discussions dating back to the mid-1980s.[42]

Shifting Power Relations

Several strands of criticism intertwine in a more contemporary critique of agricultural biotechnology's social consequences. Critics of food and agricultural biotechnology claim that policy making has been dominated by men who exhibit a decision-making style that has been the target of the feminist social movement. They note the prevalence of a viewpoint that characterizes critical attitudes as emotional or irrational and equates rational decision making with an emphasis on economics and cost-benefit-style comparison of decision options. They also believe that decision makers see nature as an object of human domination. Consistent with much of the literature in feminism, they see the domination of nature and the domination of women as themes with a common historical, intellectual, and cultural origin. Hence they argue that opposition to biotechnology and the overthrow of the existing decision-making elite for biotechnology follows from an ethical commitment to feminist philosophies of social justice. Vandana Shiva is particularly known for linking feminist ethics to the critique of the Green Revolution noted above.[43] The argument has been made as a more general postmodern critique of both agricultural and medical biotechnology by social critics such as Chaia Heller[44] and Finn Bowring.[45]

More general concerns have been raised in connection with industry's impact on publicly funded science. Released in 1990, *Biotechnology's Bitter Harvest* was one of the most influential publications to make a forceful ethical critique of food and biotechnology in a clear way. Although the report included a critique of biotechnology on environmental grounds, its primary argument was that U.S. agricultural universities were abandoning an ethical commitment to serve farmers, turning instead to the development of technology that would primarily benefit agribusiness and agricultural input firms.[46] This argument can be seen as a direct outgrowth of the issues concerning farming communities discussed earlier. Yet in directing the brunt of its criticism at the planning and conduct of publicly funded agricultural research, the authors of this report made claims with a substantially different ethical importance. Their argument connects with that of social critics who have been expressing concerns that commercial interests were having a growing influence on the conduct of science.[47]

A third strain of argument focuses again on issues relating to international development. Many of the world's most valuable plant genetic resources lie in the territory of developing countries, and many are found in land races. Land races are crop varieties that have been grown by indigenous farmers who have selected for valuable traits by a process of trial and error. Developed country plant breeders have made many advances by extracting these valuable traits from the seeds of land races. In the past, neither the indigenous farmers who grow land races nor the governments of their countries have been compensated for the use of these genetic resources. Critics have claimed that a double form of injustice occurs when these genetic resources are first taken without compensation and then sold back to developed countries in the form of seeds protected by patents or under plant breeders' rights.[48] This argument is also tied to the concern that biotechnology might hurt small farmers, but here the injury being done to them is in the form of property rights and arguably quite different from the traditional critique of social impacts due to the increasing size of farms and their industrial organization.

Ethical concerns about smallholder control over seeds predate the debate over biotechnology. Social critics have noted this issue with respect to the collection of germ plasm for conventional plant breeding.[49] Biotechnology has brought this set of concerns to the forefront of public attention in conjunction with legal debates over the patentability of genes and genetic sequences and over the status of patents and other forms of intellectual property in the TRIPS agreement, which established basic principles for adjudicating intellectual property disputes in the World Trade Organization. Defining and defending any given configuration of property rights is an inherently moral and philosophical exercise; hence these technically complex legal debates generally presume some sort of ethical framework in which arguments about what should and should not be recognized as property are mounted. Broadly, the case for recognizing the patentability of genes and gene sequence is a derivative of the case for intellectual property in general, and it is couched in utilitarian terms: in a setting of competitive markets, innovators benefit from their inventions only if they are kept secret and no competitors are able to use them. But the public benefits if the inventions are made public, and everyone can use them. So inventions (intellectual property) should be made public, but if they are made public too soon, inventors lose all incentive to innovate. Hence, the rationale for intellectual property rights, including patents and copyrights, is to give inventers an exclusive right to use or license the use of their invention, but only for a limited time, after which this right ceases to exist, thus maximizing public benefit.[50]

This basic argument has been challenged on many fronts. Some critics accept the basic utilitarian rationale for patents but question whether patents in biotechnology are really beneficial.[51] Others see the utilitarian view of patents simply as a subterfuge to allow the growth of capitalist social relations and corporate power.[52] Still others stress the view (noted above) that indigenous people who discover uses for plants and who develop germ plasm through generations of trial and error have a prior claim that vitiates this utilitarian rationale.[53] The ETC Group, a nongovernmental organization that has been active in opposing biotechnology, often links its criticisms of gene patents to the so-called Terminator gene, a biologically based means of protecting intellectual property by rendering seed infertile. Although intellectual property arguments can involve exacting technical detail when considered in a legal setting, it has proved relatively easy for critics of biotechnology to link the spread of intellectual property rights in biotechnology with the worst aspects of globalization.

These ethical issues associated with the shifting balance of power in society should be seen as distinct from concerns about the impact of technical change on farming communities. Someone who holds values that generally favor pursuit of food and agricultural biotechnology (in the belief that it will help address world hunger, perhaps) could still find fault with the way that the science agenda is being established in the era of biotechnology. One concern expressed at the grossest level is that pursuit of profit or receipt of funding from industry might influence the results of research intended to review the safety of products. More broadly, these seemingly seismic shifts in the role and nature of science, in the structure of international institutions, and in traditional ways of understanding ownership feed a pervasive concern about the general drift of social relations. Critics such as Vandana Shiva or Finn Bowring unify a broad array of medical, food-related, and legal trends to create a picture of biotechnology as a monolith that must be met with widespread popular resistance. At this point, concerns emerging out of a fairly straightforward need to anticipate unwanted consequences of biotechnology seem to blend together. Perhaps at this point, they take the shape of an intrinsic evil to be opposed simply for its own sake.

Special Arguments Pertaining to Biotechnology

The most sweeping ethical argument against biotechnology would be one that finds the manipulation of genes or cells to be either categorically forbidden or presumptively wrong. It is not clear whether members of the lay public who express ethical reservations about gene technology have such a view in mind, but it is reasonable to presume that some do. There are many ways in which such a claim might be stated. Empirical research indicates that many members of the lay public who find food or agricultural biotechnology ethically objectionable base their judgment on the view that it is *unnatural*.[54] Philosophers have called these objections to biotechnology "intrinsic objections," meaning that it is the activity of genetic manipulation itself that is wrong, not its consequences.[55]

Statements to the effect that biotechnology is unnatural convey a judgment of disapproval but do little to articulate the basis for that judgment. In one sense, all of agriculture is an unnatural activity, but we should not infer that all of agriculture is therefore of ethical concern. How would one spell out the belief that biotechnology is unnatural in a way that would form the basis for an argument against its use to develop agricultural crops or animals? How would one articulate an intrinsic objection to gene transfer that would cover its use in plants and animals, as well as human beings? A few strategies that have been attempted in the literature can be summarized.

1. *Genes and essences.* Since antiquity, people have thought of living things as having "essences" that constitute their essential being. Sociologists Dorothy Nelkin and Susan Lindee note a general cultural tendency to interpret genes as bearers of the traditional notions of essence and purpose that would achieve moral significance in some teleological conceptions of nature.[56] One view of agricultural biotechnology may see it as "tampering" with these "essences."[57] Criticisms voiced by activist Jeremy Rifkin suggest such a judgment,[58] and it is particularly associated with those who have argued that genetic engineering violates a species' telos.[59] The term *telos* is derived from the philosophy of Aristotle, where it was used to indicate a thing's guiding or final purpose, realized in the case of living organisms through the processes of growth, development, and reproduction that are characteristic of their species. It is associated with *teleology*, a philosophy of nature that seeks to explain biological processes in terms of function, purpose, and design. Although teleology does not necessarily prescribe particular ethical norms, versions of teleology that find a predetermined design in nature, often the work of a supernatural intelligence, move quickly to the ethical judgment that humans deviate from the preordained purposes of this plan at their physical and spiritual peril.

2. *Species boundaries and natural kinds.* Human cultures display a remarkable constancy with respect to the way that species boundaries are taken to reflect a kind of natural order, reflected in the linguistic tendency to build the system of meanings around natural kinds. Plants and animals visible to human senses and important for human purposes are described as

kinds, rather than as particular things not amenable to classification. Although different cultures parse the world around them in different ways, human languages tend to have equivalent kind-terms for *dog, cat, tree,* or *flower.* Henk Verhoog suggests that this tendency is evidence for an underlying system of purposes such as those discussed immediately above. He also makes the separate argument that biologists lack any special authority to redefine these terms to more faithfully reflect the scientific construal of kinds as interbreeding populations.[60] The force of this second argument is that modern biology is challenging the most basic way in which human beings have made sense of the world since antiquity—and so much the worse for modern biology.

3. *Emotional repugnance.* Genetic modification of foods causes an immediate reaction of repugnance among many. The most sophisticated philosophical statement of the ethical significance that should be associated with that reaction was made in a brief article by Leon Kass commenting on the announcement of Dolly, the sheep cloned by the Roslyn Institute in 1997. Kass's central argument is that mammalian cloning elicits a repulsive reaction from many and that this repugnance is sufficient ground to regard cloning as intrinsically wrong.[61] In making this case, Kass relies on a conservative tradition in ethics that harks back to the philosophical writings of David Hume, Adam Smith, and Edmund Burke. These philosophers believed that morality was based on sentiments of sympathy with others and that emotional attachments were a key component in any moral judgment. Although they lived and wrote in a pre-Darwinian culture, they also believed that emotional reactions such as repugnance reflect a deep-seated and culturally ingrained wisdom. Societal stability is the result of respecting these emotional reactions, and departure from them entails the risk of upheaval and dissolution. Kass's argument has since formed the basis for a similar argument against applications of recombinant technology to foods.[62]

4. *Religious arguments.* Many people clearly attach religious significance to species boundaries and question the wisdom of genetic engineering. Furthermore, many of the world's religions endorse specific injunctions against crossing species boundaries, interfering in reproductive processes, and

consuming proscribed foods. As noted already, some of the most plausible ways of understanding the view that biotechnology is unnatural or that it tampers with the natural order against the demands of morality involve appeals to divine authority. Furthermore, worldviews that construe nature as bearing specific forms of moral significance may also be considered as resting on religious foundations, especially when they involve beliefs that are not amenable to scientific characterization and measurement. A study by the Working Group of the Society, Religion and Technology Project of the Church of Scotland in 1999 (to which Donald Bruce, author of chapter 5 contributed) developed a comprehensive statement on Christian perspectives on genetic engineering applied to nonhuman species that emphasized the need to draw upon traditional religious sources and conceptions of spirituality in a concerted effort to think through the implications of genetic technology.[63] Other chapters in this volume examine some of these possibilities in greater detail, drawing upon the full spectrum of the world's religious traditions.

Most academic philosophers have rejected special concerns that have been raised about genetic engineering and have levied especially strong criticism against the claim that biotechnology is unnatural. Roger Straughan[64] and Gary Comstock[65] review a series of ways to extend the claim that gene technology is unnatural into a more substantive ethical argument for regulating or restricting crop biotechnology. They find either that the substantive issues do not pertain specifically to the use of rDNA techniques for gene transfer (that is, that they devolve into issues of general technological ethics) or that the characterization of naturalness is too sweeping and would apply to many well-accepted uses of technology. Bernard Rollin offers a similar analysis and characterizes arguments that appeal to the unnaturalness of gene transfer as an example of "bad ethics."[66] Mark Sagoff has replied to the suggestion that biotechnology is unnatural by reviewing the four ways in which John Stuart Mill found that something could be said to be natural, arguing that for the most part, no judgment against biotechnology can be maintained without also tarnishing ordinary plant breeding, if not agriculture itself.[67]

However, Robert Streiffer and Thomas Hedemann suggest that opinion research supporting the demand for labeling suggests that a majority of people have already found biotechnology to be intrinsically unacceptable and on this basis argue that political decision makers can no longer reject this sentiment in good conscience. The fact that people have faith-based views prohibiting a practice does not ordinarily provide a public basis for constraining or regulating that practice. Rather this fact establishes a prima facie obligation to respect these beliefs and to accommodate a believer's desire to act on faith-based beliefs in their daily life. Streiffer and Hedemann resist this turn in the argument, suggesting that if a sufficient number of people hold faith-based beliefs, it becomes appropriate to take whatever public action such beliefs dictate, subject to the qualification that the full range of political values must be taken into consideration when doing so. On this ground, they argue that intrinsic (or special) arguments provide a powerful basis for segregating and labeling products of biotechnology and could conceivably provide an argument for banning them altogether.[68]

Conclusion

In resisting the pro and con summarizing approach, I have tried to suggest that agricultural biotechnology has become caught up in several longstanding moral and political debates, as well as having introduced a few new wrinkles on its own. Two broad types of concern have been distinguished. First, it is possible that gene technology is of ethical concern because it poses risks to animal, environmental, and human interests, including not only individual health and safety but also economic and social considerations. Second, it is possible that the use of gene technology is itself the basis of concern, that there is something about the manipulation of living matter at the genetic level that is of ethical concern.

Notes

The literature reviewed in this chapter is but a fraction of that available. I have undoubtedly allowed my own views on biotechnology to color both my selection of sources to cite and my characterization of the views cited. Readers wishing a fuller characterization of my own views (as well as more extensive bibliography) should consult Paul B. Thompson, *Food Biotechnology in Ethical Perspective* 2nd ed. (Dordrecht, NL: Springer, 2007).

1. Jesper Lassen, Agnes Allansdottir, Miltos Liakopoulos, Arne Thing Mortensen, and Anna Olofsson, "Testing Times: The Reception of Roundup Ready Soya in Europe," in Martin W. Bauer and George Gaskell, eds., *Biotechnology: The Making of a Global Controversy* (Cambridge: Cambridge University Press, 2004), pp. 279–312.

2. L. J. Frewer, C. Howard, and R. Shepherd, "Public Concerns in the United Kingdom about General and Specific Applications of Genetic Engineering: Risk,

Benefit and Ethics," *Science, Technology and Human Values* 22 (1997), pp. 98–124.

3. Gregory E. Pence, *Designer Food: Mutant Harvest or Breadbasket of the World?* (Lanham, MD: Rowman and Littlefield, 2002).

4. Michael J. Reiss and Roger Straughan, *Improving Nature? The Science and Ethics of Genetic Engineering* (Cambridge: Cambridge University Press, 1996).

5. Gary Comstock, *Vexing Nature? On the Ethical Case against Agricultural Biotechnology* (Boston: Kluwer, 2000).

6. Hugh Lacey, *Values and Objectivity: The Current Controversy about Transgenic Crops* (Lanham, MA: Lexington Books, 2005).

7. Hans Jonas, *The Imperative of Responsibility: The Search for Ethics in a Technological Age* (Chicago: University of Chicago Press, 1984).

8. Henry I. Miller, "Agricultural Biotechnology, Law, and Food Biotechnology Regulation," in T. H. Murray and M. J. Mehlman, eds., *Encyclopedia of Ethical, Legal and Policy Issues in Biotechnology* (New York: Wiley and Sons, 2000), pp. 37–46.

9. Marc A. Saner, "Ethics as Problem and Ethics as a Solution," *International Journal of Biotechnology* 2 (2000), pp. 219–56.

10. Debra Jackson, "Labeling Products of Biotechnology: Towards Communication and Consent," *Journal of Agricultural and Environmental Ethics* 12 (2000), pp. 319–30; Robert Streiffer and Alan Rubel, "Democratic Principles and Mandatory Labeling of Genetically Engineered Food," *Public Affairs Quarterly* 18 (2004), p. 223.

11. Gary Comstock, "Ethics and Genetically Modified Foods," in Michael Ruse and David Castle, eds., *Genetically Modified Foods: Debating Biotechnology* (Amherst, NY: Prometheus Books, 2002), pp. 88–107.

12. David Andow, Henry Daniell, Paul Gepts, Kendall Lamkey, Emerson Nafzinger, and Dennis Strayer, *A Growing Concern: Protecting the Food Supply in an Era of Pharmaceutical and Industrial Crops* (Washington, DC: Union of Concerned Scientists, 2004).

13. Paul B. Thompson, "Agricultural Biotechnology and the Rhetoric of Risk: Some Conceptual Issues," *The Environmental Professional* 9 (1987), pp. 316–26.

14. Jack Doyle, Altered Harvest: Agriculture, Genetics and the Future of the World's Food Supply (New York: Viking, 1985).

15. Jane Rissler and Margaret Mellon, *The Ecological Risks of Engineered Crops* (Cambridge, MA: MIT Press, 1995); Sheldon Krimsky and Roger Wrubel, *Agricultural Biotechnology and the Environment: Science, Policy and Social Issues* (Urbana: University of Illinois Press, 1996).

16. Laura Westra, *Living in Integrity* (Totowa, NJ: Rowman and Littlefield, 1997).

17. John Durant, Martin W. Bauer, and George Gaskell, *Biotechnology in the Public Sphere* (London: Science Museum, 1998).

18. L. L. Wolfenbarger and P. R. Phifer, "The Ecological Risks and Benefits of Genetically Engineered Plants," *Science* 290 (2000), pp. 2088–93.

19. Donald Bruce, "Contamination, Crop Trails and Compatibility," *Journal of Agricultural and Environmental Ethics* 16 (2003), pp. 595–604.

20. Deborah Rudacille, *The Scalpel and the Butterfly: The War between Animal Research and Animal Protection* (New York: Farrar, Straus, and Giroux, 2000).

21. Immanuel Kant and Louis Infield, trans., *Lectures on Ethics* (New York: Harper and Row, 1963), pp. 239–41.

22. Richard Sorabji, *Animal Minds and Human Morals* (Ithaca, NY: Cornell University Press, 1993).

23. Aquinas is quoted in Tom Regan and Peter Singer, eds., *Animal Rights and Human Obligations*, 2nd ed. (Englewood Cliffs, NJ: Prentice-Hall, 1989), p. 9.

24. Bentham is excerpted in Regan and Singer, p. 26.

25. Peter Singer, *Animal Liberation* (New York: HarperCollins, 2002).

26. Tom Regan, *Animal Rights and Human Wrongs: An Introduction to Moral Philosophy* (Totowa, NJ: Rowman and Littlefield, 2003).

27. James M. Jasper and Dorothy Nelkin, *The Animal Rights Crusade: The Growth of a Moral Protest* (New York: Maxwell Macmillan International, 1992).

28. T. B. Mepham, R. D. Combes, M. Balls, O. Barbieri, H. J. Blokhuis, P. Costa, R. E. Crilly, T. de Cock Buning, V. C. Delpire, M. J. O'Hare, L.-M. Houdebine, C. F. van Kreijl, M. van der Meer, C. A. Reinhardt, E. Wolf, and A-M. van Zeller. ECVAM Workshop Report 28: The Use of Transgenic Animals in the European Union, *Alternatives to Laboratory Animals* 26 (1998), pp. 21–43.

29. Paul Sparks, Richard Shepherd, and Lynn Frewer, "Assessing and Structuring Attitudes towards the Use of Gene Technology in Food Production: The Role of Perceived Ethical Obligation," *Journal of Basic and Applied Social Psychology* 16 (1995), pp. 267–85.

30. Michael W. Fox, "Transgenic Animals: Ethical and Animal Welfare Concerns," in P. Wheale and R. McNally, eds., *The Bio-Revolution: Cornucopia or Pandora's Box* (London: Pluto, 1990), pp. 31–54; Richard Ryder, "Animal Genetic Engineering and Human Progress," in P. Wheale and R. McNally, eds., *Animal Genetic Engineering: Of Pigs, Oncomice and Men* (London: Pluto, 1995), pp. 1–16; Andrew Linzey, *Animal Theology* (Urbana: University of Illinois Press, 1995).

31. Bernard Rollin, "Bad Ethics, Good Ethics and the Genetic Engineering of Animals in Agriculture," *Journal of Animal Science* 74 (1996), pp. 535–41; Gary Varner, "Cloning, Overview of Animal Cloning," in T. Murray and M. Mehlman, eds., *Encyclopedia of Ethical, Legal and Policy Issues in Biotechnology* (New York: Wiley and Sons, 2000), pp. 139–48.

32. Bernard Rollin, *The Frankenstein Syndrome: Ethical and Social Issues in the Genetic Engineering of Animals* (New York: Cambridge University Press, 1995).

33. Ibid.

34. Douglas Powell and William Leiss, *Mad Cows and Mothers' Milk: The Perils of Poor Risk Communication* (Montreal and Kingston: McGill-Queens University Press, 1997).

35. Rollin, *The Frankenstein Syndrome*.

36. P.Sandøe, B. L. Nielsen, L. G. Christensen, and P. Sørensen, "Staying Good While Playing God: The Ethics of Breeding Farm Animals," *Animal Welfare* 8 (1999), pp. 313–28.

37. Richard Kirkendall, "The Central Theme of American Agricultural History," *Agriculture and Human Values* 1, no. 2 (1984), pp. 6–8.

38. Joel Schor, *The Evolution and Development of Biotechnology: A Revolutionary Force in American Agriculture*. Economic Research Service, U.S. Department of Agriculture, Staff Report no. AGES 9424 (1994).

39. Nuffield Council on Bioethics, "Genetically Modified Crops: The Ethical and Social Issues," 1999; Nuffield Council on Bioethics, "The Use of Genetically

Modified Crops in Developing Countries," 2003. Retrieved on May 31, 2005 from http://www.nuffieldbioethics.org.

40. Norman Borlaug, "Ending World Hunger: The Promise of Biotechnology and the Threat of Antiscience Zealotry," *Plant Physiology* 124 (2001), pp. 487–90.

41. Kenneth A Dahlberg, *Beyond the Green Revolution: The Ecology and Politics of Global Agricultural Development* (New York: Plenum, 1979).

42. Jack Kloppenburg Jr., "The Social Impacts of Biogenetic Technology in Agriculture: Past and Future," in G. M. Berardi and C. C. Geisler, eds., *The Social Consequences and Challenges of New Agricultural Technologies* (Boulder, CO: Westview, 1984), pp. 291–323; Martin Kenney and Frederick Buttel, "Biotechnology: Prospects and Dilemmas for Third World Development," *Development and Change* 16 (1985), pp. 61–91; Henk Hobbelink, *Biotechnology and the Future of World Agriculture* (London: Zed Books, 1991).

43. Shiva's writings are extensive. For a representative sample, see Vandana Shiva, *Stolen Harvest: The Hijacking of the Global Food Supply* (Cambridge, MA: South End, 2000). For a more philosophically sophisticated sample, see Vandana Shiva, "Epilogue: Beyond Reductionism," in Vandana Shiva and Ingunn Moser, eds., *Biopolitics: A Feminist and Ecological Reader on Biotechnology* (London: Zed Books, 1995), pp. 267–84.

44. Chaia Heller, "McDonald's, MTV and Monsanto: Resisting Biotechnology in the Age of Information Capital," in Brian Tokar, ed., *Redesigning Life? The Worldwide Challenge to Genetic Engineering* (London: Zed Books, 2001), pp. 405–19.

45. Finn Bowring, *Science, Seeds and Cyborgs: Biotechnology and the Appropriation of Life* (London: Verso, 2003).

46. Rebecca Goldberg, Jane Rissler, Hope Shand, and Chuck Hassebrook. *Biotechnology's Bitter Harvest: Herbicide Tolerant Crops and the Threat to Sustainable Agriculture* (Washington, D C: Biotechnology Working Group, 1990).

47. Sheldon Krimsky, *Biotechnics in Society: The Rise of Industrial Genetics* (New York: Praeger, 1991); Lawrence Busch, William B. Lacy, Jeffery Burkhardt, and Linda R. Lacy, *Plants, Power and Profit: Social, Economic, and Ethical Consequences of the New Biotechnologies* (Cambridge, MA, and Oxford:Blackwell, 1991).

48. Hope Shand, "Gene Giants: Understanding the "Life Industry," in Brian Tokar, ed. *Designing Life? The Worldwide Resistance to Genetic Engineering* (London: Zed Books, 2001), pp. 222–37; David Magnus, "Intellectual Property and Agricultural Biotechnology: Bioprospecting or Biopiracy?" in David Magnus and Glenn McGee, eds., *Who Owns Life?* (Amherst, NY: Prometheus Books, 2002), pp. 265–76.

49. Calestus Juma, *The Gene Hunters: Biotechnology and the Scramble for Seeds* (Princeton, NJ: Princeton University Press, 1988).

50. Baruch A. Brody, "An Evaluation of the Ethical Arguments Commonly Raised against the Patenting of Transgenic Animals," in William Lesser, ed., *Animal Patents: The Legal, Economic and Social Issues* (New York: Stockton, 1989).

51. Ned Hettinger, "Patenting Life: Biotechnology, Intellectual Property, and Environmental Ethics," *Boston College Environmental Affairs Law Review* 22 (1995), pp. 267–75.

52. Beth Burrows, "Patents, Ethics and Spin," in Brian Tokar, ed., *Designing Life? The Worldwide Challenge to Genetic Engineering* (London: Zed Books, 2001), pp. 238–51.

53. Victoria Tauli-Corpuz, "Biotechnology and Indigenous Peoples," in Brian Tokar, ed., *Designing Life? The Worldwide Challenge to Genetic Engineering* (London: Zed Books, 2001), pp. 252–70.

54. Wolfgang Wagner, Nicole Kronberger, Nick Allum, Suzanne De Cheveigné, Carmen Diego, George Gaskell, Marcus Heinßen, Cees Midden, Marianne Ødegaard, Susanna Öhman, Bianca Rizzo, Timo Rusanen, and Angelici Stathopoulou, "Pandora's Genes: Images of Genes and Nature," in Martin W. Bauer and George Gaskell, eds. *Biotechnology: The Making of a Global Controversy* (Cambridge: Cambridge University Press, 2002), pp. 244–76.

55. Marc A. Saner, "Real and Metaphorical Moral Limits in the Biotech Debate," *Nature Biotechnology* 19 (2001), p. 609.

56. Dorothy Nelkin and M. Susan Lindee, *The DNA Mystique: The Gene as a Cultural Icon* (New York: Freeman, 1995).

57. Jochen Bockmühl, "A Goethean View of Plants: Unconventional Approaches," in D. Heaf and J. Wirz, eds., *Intrinsic Value and Integrity of Plants in the Context of Genetic Engineering* (Llanystumdwy, UK: International Forum for Genetic Engineering, 2001), pp. 26–31.

58. Jeremy Rifkin, *Declaration of a Heretic* (Boston and London: Routledge and Kegan Paul, 1985); "Farm Animals and the Biotechnology Revolution," in Peter Wheale and Ruth McNally, eds., *Animal Genetic Engineering: Of Pigs, Oncomice and Men* (London: Pluto, 1995), pp. 19–38.

59. Henk Verhoog, "The Concept of Intrinsic Value and Transgenic Animals," *Journal of Agricultural and Environmental Ethics* 5 (1992):147–60; "Biotechnology and Ethics," in T. Brante, S. Fuller, and W. Lynch, eds., *Controversial Science* (New York: State University of New York Press, 1993), pp. 83–106.

60. Ibid.

61. Leon Kass, "The Wisdom of Repugnance," *The New Republic*. June 2, 1997, pp. 17–26.

62. Mary Midgley, "Biotechnology and Monstrosity," *The Hastings Center Report* 30, no. 5 (2000), pp. 7–15.

63. Donald and Ann Bruce, eds., Engineering Genesis: The Ethics of Genetic Engineering in Non-Human Species (London: Earthscan, 1998).

64. Roger Straughn, "Ethical Aspects of Crop Biotechnology," in T. B. Mepham, G. A. Tucker, and J. Wiseman, eds. *Issues in Agricultural Bioethics* (Nottingham: University of Nottingham Press, 1995), pp. 163–76.

65. Gary Comstock, "Is It Unnatural to Genetically Engineer Plants?" *Weed Science* 46 (1998), pp. 647–51.

66. Rollin, "Bad Ethics, Good Ethics."

67. Mark Sagoff, "Biotechnology and the Natural," *Philosophy and Public Policy Quarterly* 21 (2001), pp. 1–5.

68. Robert Streiffer and Thomas Hedemann, "The Political Import of Intrinsic Objections to Genetically Engineered Food," *Journal of Agricultural and Environmental Ethics* 18 (2005), pp. 191–210.

3

Does Vegetarianism Preclude Eating GM Foods?

Lyne Létourneau

Introduction

Throughout the ages, meat has remained an ambiguous food product, giving rise to ambivalent feelings of appreciation and abhorrence.[1] For instance, whereas eating meat marked for several centuries one's privileged status within society, eschewing flesh consumption constituted for others a means to spiritual elevation.[2] It is interesting to note, however, that, across all human societies, food taboos possess the remarkable particularity of almost always involving substances of animal origin.[3]

An ancient example still flourishing today, and dating back thousands of years, is vegetarianism. The term *vegetarianism* applies indiscriminately to a wide range of dietary practices that avoid foods from animal sources with varying degrees of restriction. It covers a spectrum of diets that extend from "semivegetarianism," or the avoidance of meat, particularly red meat, but not of fish, seafood, eggs, and dairy products, to "veganism," which excludes the consumption of all animal flesh and animal products. Between these two extremes, one finds vegetarians who do not eat meat, including fish and seafood, but who consume eggs and dairy products ("lacto-ovo vegetarianism"), as well as vegetarians who avoid either eggs or dairy products in addition to meat, fish, and seafood (correspondingly "lacto-vegetarianism" and "ovo-vegetarianism").[4] Despite these differences, the common characteristic of all vegetarian diets is that they are plant-based, resting mainly, or exclusively, on the consumption of grains, vegetables, fruits, legumes, seeds, and nuts.[5]

Although vegetarian dietary practices now enjoy greater accessibility through the establishment of a niche market for vegetarian food, and although vegetarianism is becoming ever more prevalent in industrialized countries, surveys in the United Kingdom situate the proportion of vegetarians at approximately 5 percent of the population.[6] In the United States and Canada, vegetarians constitute between 1 percent and 2.5 percent of the population.[7] These figures are not considerable and justify looking at vegetarianism as a marginal diet.

Yet, as subsidiary as it may be, advances in the genetic engineering of foods might constitute a potential threat to the vegetarian diet. To be sure, as explained in Abraham's chapter, the techniques of cell manipulation and transgene expression permit the transfer of genes between species, including between animals and plants. With the improvement of DNA technologies, the sequencing of ever more genomes, increasing knowledge in genomics and proteomics, the international expansion of bioprospecting, and the development of the ability to manipulate complex genetic traits through the advent of artificial chromosomes, the door might open up to the engineering of plants destined for human consumption with added DNA from animal origin.

For now, such likelihood belongs to an uncertain future. However, if such food products were to reach market, would they be acceptable to vegetarians? Or would the restrictions imposed by vegetarianism be violated by genetically modified plants containing either one or many transgenes coming from animal sources? These are important questions because the rapid adoption of new commercially valuable transgenic plant breeds might hinder access to plant-based food products that satisfy the requirements of vegetarian dietary practices.

In the following pages, I will examine how adherence to vegetarianism affects the acceptance of otherwise suitable food products containing transgenes from animal origin. First, I will present a broad outline of the vegetarian ideology, which "provides both a critique of meat eating and the vision of a vegetarian world."[8] Second, arguing on the backdrop of the vegetarian worldview, I will circumscribe the extent of the menace posed to vegetarianism by genetically modified plants containing added DNA from animal sources. In the course of my discussion, I will draw from both existing literature and data recently gathered from two focus groups on vegetarianism and genetically modified foods.[9]

The Vegetarian Worldview

So far, vegetarianism has been described as a food behavior, a dietary preference. However, vegetarianism also refers to an ideology, that is, "a construct that contains . . . value associations and guides behavioural decisions."[10] Indeed, eating is a complex social fact. For it brings into play an intermingled network of cultural, spiritual, political, material, and ethical considerations. As stated by socio-anthropologist Jean-Pierre Poulain: "*On ne se nourrit pas que de nutriments.*"[11] To be sure, food choices are not made on a strictly nutritional basis. Rather, foods are invested with symbols and reflect social representations. Therefore, when we eat foods, we are also consuming meanings, that is, the products of pervasive and ongoing social construct processes.[12] What is more, it is generally recognized that one's choice of food products provides an efficient way to express one's identity, one's own system of values and beliefs.[13] Particularly, in a world wherein conventional references are increasingly confused, dietary practices act as powerful identity markers.[14]

According to sociologist Donna Maurer, "[f]or many people . . . being a vegetarian means more than following a set of dietary proscriptions—it is a way of life."[15] For these people, vegetarianism constitutes a means to express their identity as individuals, to affirm their values and beliefs. As explained by psychologists Marjaana Lindeman and Minna Sirelius, "[A]lthough vegetarianism on the surface may seem to purport a lifestyle constricted by nutritional guidelines; this normative aspect of a vegetarian lifestyle may not be dominant in vegetarianism. The restrictions seem hence not to be dictated by externally imposed norms, but can instead be viewed as freely chosen requirements for exercising one's view of world, of doing something concrete for those issues . . . that one cherishes."[16]

Vegetarians as a group do not form a homogenous community. Be that as it may, they share a fundamental concern about meat avoidance. What is more, their choice of a plant-based diet rests on a set of reasons that, although they are not shared universally among vegetarians, are nonetheless relatively fixed. These reasons state the case for vegetarianism and, taken together, propose an ideology, or worldview, wherein vegetarians' particular conception, or ideal, of what should be the relationships among humans, animals, nature, and society as a whole is defined.[17]

Maurer, who has studied vegetarianism as a social movement, identifies three basic tenets of vegetarians' system of values and beliefs: (i) compassion for all living beings, which includes social justice; (ii) promotion of health and vitality; and (iii) concern for the environment. As she underscores, "[a]dvocates and movement leaders sometimes debate the finer points of vegetarianism (for example, does honey harvesting cause bees to suffer?), but—because the ideology reflects a long-standing interconnected system of beliefs and ideas to which longtime vegetarians and newcomers alike refer for justification of their practices—they rarely contest its basic tenets."[18]

Among the basic tenets of vegetarianism, the health benefits to be gained from a meatless diet are reported as the most often cited rationale for espousing a vegetarian diet.[19] Human studies on nutrition and disease indicate that "the elimination of meat from the diet is indeed one of the most powerful determinants of human health,"[20] playing a role in the prevention of obesity, cardiovascular disease, stroke, hypertension, type 2 diabetes, osteoporosis, colon cancer, breast cancer, etc.[21] Generally speaking, individuals who adopt a vegetarian diet on the basis of personal health concerns are referred to as "health vegetarians."

"Ethical vegetarians," on the other hand, embrace a plant-based diet because they hold the belief that eating meat is morally wrong.[22] In their anthology devoted to vegetarianism, philosophers Kerry S. Walters and Lisa Portmess[23] identify the argument from moral concern over animals as the foundation of ethical vegetarianism.[24]

Asserting the kinship of all living things, Pythagoras is believed to have offered the first defence of vegetarianism based on this argument.[25] His objections to eating meat were tied to his belief in the transmigration of souls. According to him, "the soul was immortal and could be endlessly transformed into other living creatures."[26] Consequently, "[t]o kill and eat any living creatures, whether they be bird, reptile or fish, was to murder one's cousins and eat their flesh."[27] For Pythagoras, "the consumption of a food animal may in fact be the devouring of a human soul."[28] Since Pythagoras, the argument from moral concern over animals has known many articulations.[29] These either stem from the calling into question of our instrumental use and killing of animals for food or arise from a disquiet regarding our abusive treatment of them.

Having crossed centuries, the argument from moral concern over animals is supplemented by contemporary defences of ethical vegetarianism that call attention to the ecological devastation caused, directly or indirectly, by factory farming and intensive agricultural practices. It is argued that "meat production depletes water supplies, forests, and fossil fuel energy and causes soil erosion,"[30] thereby decreasing the ability of ecosystems to maintain and regenerate themselves. Raising animals for food is thus considered to be incompatible with environment sustainability.

Meat eating is also perceived as contributing to the perpetuation of global food inequalities, a line of reasoning that is related to the basic tenet of social justice as an expression of compassion for all living beings. Converting grain and soy to meat is regarded as a very inefficient and wasteful method of producing food because "an acre of cereals can produce *five times* more protein than an acre devoted to meat production; legumes (peas, beans, lentils) can produce *ten times* more; leafy vegetables *fifteen times* more."[31] In his book *Food for a Future*, Jon Wynne-Tyson states the argument as follows: "About four-fifths of the world's agricultural land is used for feeding animals and only about one-fifth for feeding man directly. . . . We read in our newspapers about the starving and under-fed millions, and all the time we are feeding to meat-producing animals the very crops that could more than eradicate world food shortage; also, we are importing from starving nations large quantities of grain and other foods that are then fed to our animals instead of to the populations who produced them."[32]

Under such a view, social justice would require eliminating the consumption of meat in order to meet the dietary needs and requirements of the entire human population.

The argument from social justice finds another expression in the connection that is made between ethical vegetarianism and feminism. As a "reflection of patriarchal power structures that are inherently exploitative," a causal relationship is established between the consumption of animal flesh and the oppression of women.[33]

Except for this last argument, which appears to be restricted still to philosophical circles, empirical studies confirm that, in addition to health, vegetarian dietary practices are adopted for all of the above-mentioned reasons: disquiet with the treatment of farm animals and the rejection of killing, a preoccupation with the protection of the environment, and an aspiration toward social justice.[34]

The opinions expressed by the participants in our two focus groups emphasized moral consideration for animals and environmental protection as their main basis for vegetarianism. To illustrate, consider the following statements:

> For me, any of the foods that . . . are generated due to any form of cruelty [are] like a complete "no."[35]
> [I]t is a matter of conscience and dietary preference, in the sense that I just do not want to consume anything that has undergone hardship or torture, you know.[36]

> So whether there was suffering involved or not, the very fact that I'm taking a life, where I could definitely . . . be avoiding that, is what I subscribe to.[37]

> I have a definite preference for organic because part of my reason for being vegan is an environmental reason . . . [38]

> Some people mentioned that we're at the point where we've screwed up things a lot, and to get to a point where we're eating food in a sustainable manner, that leaves things the way they are, for the future generations . . . [we need] a large overhaul. To get a process of reversal, to get to a point where we are in a

state of sustainability. . . . [And sustainability relates directly to veganism or vegetarianism] a hundred percent, yes.[39]

In addition, some of the focus group participants insisted on the dis-connection between consumers and food sources. The unease generated by this perceived distance motivated their preference for plant-based foods that are produced locally, as illustrated by the following statement: "[W]e are losing our connection . . . with our food and the people who produce it."[40] Such an argument, however, does not support the case for vegetarianism in its imperative to abstain from meat consumption,[41] for it is associated with a contemporary critique of food production as a whole, including both animal and plant-based products.

To be sure, profound mutations occurred within the food production system throughout the last 150 years. These transformations deeply altered our relation to foods, creating an increasing distance between consumers and foods.[42] Notably, because urbanization removed food products from their rural production sites, it conferred to them a status of merchandise and blurred their deep-rootedness in the natural environment. The industrial-ization of food production also further severed the link between food and nature. Such an effect is exemplified by the intensive breeding and keep-ing of animals for farming purposes, which contributed to the reification of animals destined for human consumption by reducing them to raw material. Large-scale distribution, the development of a thriving food transformation industry, and the impact of food marketing strategies deserve to be men-tioned too as factors of disconnection between consumers and foods.[43]

In summary, many reasons substantiate the case for vegetarianism. From personal health benefits to global food inequalities, including moral concern over animals and environmental protection, the arguments put forward count as many constitutive elements of the vegetarian worldview. Expressing an ideal state of relationships between humans and their social and physical environment, vegetarians' system of values and beliefs empha-size prudence, well-being, compassion, nonviolence, sharing, and respect as key guides for action. Through their dietary choices, vegetarians live in accordance with their view of world.

The likelihood, therefore, that access to food products that are appro-priate for vegetarians be limited through genetic engineering goes far beyond the logistics of procuring food and dietary preferences. For the effect of such constraint relates to the free pursuit of one's self. In a liberal and democratic society, such a situation cannot be indifferent. Yet, what is the extent of the threat posed to vegetarianism by genetically modified plants containing added DNA from animal sources?

The Threat to Vegetarianism

According to bioethicist Arthur Caplan, the debate on genetic engineering is closed. Speaking about genetically modified animals, he made the following statement: "I don't have any problem engineering animals. Ethically, we've answered that."[44] Such a strong assertion, however, may be doubted because, as shown in Thompson's chapter, not only does the controversy about the application in the agricultural setting of the techniques of cell manipulations and transgene expression raise concerns associated with the specific characteristics of the technology, but it also correlates with issues of food production that involve a reflection on the ethics of technological progress. Thompson points out, as a matter of fact, that "agricultural biotechnology has become caught up in several longstanding moral and political debates." We are thus very far from any consensus on the acceptability of GM foods, and the debate continues.

In discussing whether plant-based food products with added DNA from animal origin threaten the vegetarian lifestyle, the above point is important to remember because participants in the focus groups tended to make their arguments not so much on the basis of the case for vegetarianism, but from the broader perspective of the controversy that still vividly surrounds genetic engineering of food products, invoking reasons such as respect for nature, the unnaturalness of genetic engineering, its unknown long-term consequences for the environment and human health, mistrust toward industry and government, and GM foods being an unacceptable "band-aid" solution to world hunger. These broader justifications, however, are relevant for our present purposes only to the extent that they actually fit within the vegetarian worldview. What is more, in reaching a conclusion, their invocation does not spare the need to delineate the implications of the arguments supporting vegetarians' choice of a plant-based diet with respect to genetically modified plants containing added DNA from animal sources. Both sets of reasons must be dealt with.

I will begin with the argument from moral concern over animals. As a general rule, each use of animals—whether this is for research, companionship, the production of food, entertainment, or sport—raises three basic moral questions. The first one is whether this particular form of animal use is morally acceptable. If it is, the second question asks whether any limitations should be placed on this use. Finally, the third question relates to the standards of treatment of animals. In other words, it concerns the sort of treatment that should be accorded to animals in the context of this particular use of animals.[45] The first two questions refer to a first level of moral reflection, that of animal use per se. The third one pertains to a second level of moral reflection, that of the actual treatment of the animals used. Taken together, both levels of moral reflection set a structure of analysis that envisages each use of animals as, first, "a framework activity" together, if morally acceptable, with a set of standards of treatment.

In the debate over eating meat, the rationales put forward in favor of vegetarianism address both levels of moral reflection. As indicated in the first section, the contentions against the consumption of meat either stem from the calling into question of our instrumental use and killing of animals for food or arise from a disquiet regarding our abusive treatment of them. When considering the purport of the argument from moral concern over animals with respect to genetic engineering, the level upon which moral reflection is directed must be taken into account.

Broadly stated, at the first level of moral reflection, opposition to meat rests on the contention that it is morally wrong to treat animals as resources to be exploited for humans' benefit or, in other words, as means to achieve our ends. Animals are believed to possess moral status,[46] and their raising and killing for the production of food is viewed as a violation of their most basic interests in life and the avoidance of suffering. Animal farming therefore is looked upon as disregarding the moral status of animals. Set within the context of the use of cell manipulations and transgene expression techniques, the question becomes whether the transfer of genetic information from an animal to a plant also infringes on the moral status of animals. The following quote from one focus group participant reflects the issue at hand: "[E]ven humans have to sign a consent form to give, to donate organs and this kind of thing, and I think that the way that we treat animals as we sort of subjugate them…—like they don't have a voice to speak up, and I just think that in a way we almost violate them."[47]

This is a complex philosophical matter that cannot be fully explored here—especially considering the fact that it is interrelated with a similar interrogation associated with the use of human genetic material. Still, a few observations can be made on the backdrop of some of the opinions mentioned during the focus groups.

One participant expressed the view that taking a gene from an animal violated its "essence," denigrating its "integrity."[48] Such a claim must be distinguished from an argument based on the moral status of animals. For it usually refers to *species* integrity, not *individual* integrity.

Frequently invoked in the debate on genetic engineering, the argument from species integrity rests on the assumptions that species are definite entities, characterized by unique and permanent "essences" that constitute their essential being and that species have the specific function of setting apart the various groups of living beings. According to the argument, species possess explicit "boundaries" that are defined, among other things, by the genes that make up their genomes. To be sure, genes are seen as the bearers of the "essences" that define each species. What is more, the argument maintains that the biological units that species represent possess value in themselves, or intrinsic value. Therefore, the integrity of species, or, more specifically, of species "boundaries," must be respected, including in their genetic component.

Applied to the case at hand, the argument from species integrity leads to the conclusion that genetically modifying plants with added DNA from animal origin violates species "boundaries" through the genetic modification of the plants' genomes. Such a conclusion reverberates the aforementioned statement from our focus group participant. However, respect for the intrinsic value of species is not the same as respect for the moral status of animals—which in any case may or may not recognize the intrinsic value of animals.[49] The former shares much closer links with environmental ethics than with animal ethics. Thus, severed from any connection to the argument from moral concern over animals, the argument from species integrity finds little relevance at the first level of a moral reflection concerning animal use per se in the context of genetic engineering.

Another focus group participant affirmed that eating a plant-based food product containing transgenes from animal origin was tantamount to eating meat. As stated, "crossing a fish with a tomato, no, the tomato's not a fish, but it has a fish gene in it, and I'm not a vegetarian if I'm eating that tomato."[50]Underlying this belief is a conception of animal identity that is closely associated with an animal's genome. According to such a conception, cross-species engineering is unacceptable because it does not take into account the importance of genes in the definition of animal identity. Rather, genetic engineering conveys a reductionist conception of animals in keeping with which genetic information is instrumental and does not constitute an integral part of an animal's identity. Genetic engineering also carries a dualist conception of animals that separates and opposes genetic information to the other constitutive elements of animal identity, the genome being subordinate to these other elements. However, according to this focus group participant's conception of animal identity, genes do matter. Therefore, putting animal genes into plants incorporates them as a constitutive element of an animal's identity. It thus follows that eating plant-based food products containing transgenes from animal origin may be considered under such a view as equivalent to eating an animal—and hence animal flesh.

Yet, such a conception of animal identity is distinct from the concept of moral status, since the issue of what it means to be a "fish," a "dog," or a "horse" is separate from the question of what moral weight should be accorded to animals' interests. Although the former definition will likely affect one's delineation of the scope of application of the concept of moral status, it still concerns a different matter. That being the case, the argument for animal identity does not exemplify, at the first level of moral reflection, any implication of the argument from moral concern over animals in the context of genetic engineering.

What line of reasoning might lead to such an inference? As intimated here, one approach available to vegetarians who oppose eating meat based on their calling into question of animal use per se underscores the possibility that research associated with the development of GM food products might violate the moral status of animals, thereby precluding the consumption of these products upon reaching market: "[W]ithout doubt, animals are used inappropriately in the testing and research process, and for me . . . [this is] unacceptable, end of story."[51] "I think as a vegan it's totally unacceptable to have an animal's product to modify something that I would eat, totally unacceptable. How is that animal treated? Was it killed to get that gene?"[52]

Alternatively, one could maintain that using DNA from animal origin, while not constituting as such a negation of the moral status of animals, does nonetheless contribute to the spread of an instrumental view of animals. Such an argument, however, presupposes a conception of animal identity that is closely associated with an animal's genome—as described earlier. Therefore, it might not be accepted by all vegetarians. Moreover, short of being universally shared among vegetarians, it remains to be demonstrated empirically whether such a notion of animal identity is sufficiently widespread to be regarded as forming part of the vegetarian worldview.

Moving from the first stage of moral reflection to the second level, it is useful at this point to cite one focus group participant, who summarizes the distinction between the two levels: "There are some people that believe any use of animals is inherently immoral. I don't subscribe to that . . . My concern is animal welfare and animal suffering, so I think it is acceptable to use animals provided their suffering does not result from that."[53]

Since the 1964 publication of Ruth Harrison's book *Animal Machines*,[54] the intensive conditions under which farm animals are raised have been criticized widely. To illustrate, from the battery cage system, in which laying hens spend their lives confined in small wire cages with several other hens and cannot exercise their natural behaviors, to the production of "white veal" and the keeping of sows in stalls with very little space to turn round and lie down comfortably, the attacks put forward against intensive animal husbandry practices have been extensive.[55] At the end of the 1990s, criticisms even intensified as the attention of the animal protection movement turned from the use of animals in biomedical research toward the use of animals for the production of food.

Generally speaking, second-level opposition to the breeding, raising, and killing of animals for food is based on the view that animal agriculture involves a considerable amount of suffering inflicted on animals through inhumane and abusive treatment. This argument, however, is irrelevant to the genetic engineering of plant-based food products containing transgenes from animal origin, since the negative assessment that it articulates, and that substantiates vegetarians' abstention from meat, is specific to animal farming and does not extend to any other activity.

Nevertheless, the argument exemplifies compassion for animals as part of the vegetarian worldview. Therefore, although neither vegetarian nor animal protection organizations are coterminous,[56] it is likely that the infliction of animal suffering in the course of the overall process leading to the development of genetically modified plants with added DNA from animal sources (i.e., bioprospecting, identification of the functions of animal genes, gene sampling, etc.) would be condemned by vegetarians as a contravention of one of the three basic tenets of vegetarianism, thereby precluding their consumption of these food products.

In the absence of animal suffering, however, disapproval would not be warranted unless other relevant justifications are put forward. Could such justifications be found in the arguments from environmental protection, global food inequalities, and personal health benefits, which also support the case against the use of animals for the production of food?

Like the second level of the argument from moral concern over animals, the arguments from environmental protection, global food inequalities, and personal health benefits are *activity-specific* in the sense that they aim to criticize animal farming and the consumption of food products derived from animal sources. Now, genetic engineering of plants with added DNA from animal origin is a distinct activity from animal farming. That being the case, it does not fall within the scope of application of these arguments. The latter provide no basis against including in one's diet genetically modified plants containing added DNA from animal sources.

Still, as constitutive of the vegetarian worldview, the arguments from environmental protection, global food inequalities, and personal health benefits express vegetarians' care for the environment, consideration for social justice, and sensitivity to human health issues. They point to interconnectedness between vegetarians' system of values and beliefs and achievement of the wider societal goals born by these aspirations. Of course, the affinity of vegetarianism for other social movements would require much further investigation and should be confirmed through empirical research.[57] Nevertheless, one should expect vegetarians to be supportive of many of the arguments put forward against the genetic engineering of plants for human food production that rest on a set of values and beliefs similar to theirs—including the following:

- It will lead to loss of genetic diversity and perturbation of ecosystems.
- It will widen the gap between developing countries and industrialized nations.
- It will reinforce concentration of power in the hands of industry.
- It will create unknown risks of toxicity, bioactivity, and allergenicity for human health.

Indeed, as illustrated below, the opinions expressed by the participants in our two focus groups resonated greatly with some of the above claims:

One of the reasons that, I mean, I would be against [the idea of a gene or genes from an animal being introduced into a plant product] is, it's just the fact that I feel that nature's taken millions of years to develop something, and if man interferes and just messes with it, just with our short-sighted thinking we can actually cause more damage than we can actually imagine at that moment. So that's one of the biggest reasons, I mean, that I'm kind of against genetic modification.[58]

[I]f you took an evolutionary standpoint, nature kept these gene pools separate probably for a reason, and if we tinker with it then who knows what the consequences are going to be.[59]

[I]t's not just like something like a brand of shoes that, well, if that doesn't work then we can get another type or make our own, whatever. This is food. . . . Something that we depend on to survive. So even forgetting about the environment for a moment here . . . our own survival can be threatened.[60]

[I]f all of the GMO corn crops fail, what are we going to do? If all of the, you know, soy crops fail, what are we going to do?[61]

[L]ike okay, you have a gene from a fish that ends up in a tomato but suppose that that same chemical sequence could be synthesized, would it make a difference? You know, I guess for me it's the broader issues about the environment and the ultimate effect on your health.[62]

I think I'd say . . . not knowing what the repercussions are, and for me that's a big point, there has been no long-term testing. We are guinea pigs right now. I didn't consent to that, to the fact this science is happening without long-term testing and the fact that it's being released into the market place. This is a big concern, and to see how humans will react to this—the increase in allergic reactions. I think that there's a lot of things that can happen with the human body but perhaps we're not able to pinpoint it, because we don't have long-term testing.[63]

There's potential for abuse, the corporations that are profiting from this that can trademark the type of foods.[64]

[T]here's lots of issues around ownership, like the financial repercussions that this is having on producers and . . . we have a small group of multinationals that control our food source.[65]

Therefore, although the arguments from environmental protection, global food inequalities, and personal health benefits do not justify concluding that genetically modified plants with added DNA from animal origin pose a threat to the vegetarian lifestyle, they do reflect a worldview in relation to which aspects of the genetic engineering of plants appear to be problematic. The opposition, however, is not directed specifically toward plant-based food products containing transgenes from animal sources but toward all GM foods derived from plants. These food products are deemed unacceptable because the application in the agricultural context of the techniques of cell manipulations and transgene expression is inconsistent as a whole with vegetarians' view of the world.

Of course, were plant-based GM food products to be proved to be without risk for the environment, inoffensive to social justice, and safe for humans, there would remain in accordance with the vegetarian worldview no obstacle to their consumption.

Although limited to genetically modified plants with added DNA from animal origin, a similar conclusion was reached with respect to the second level of the argument from moral concern over animals. As for the first stage of moral reflection associated with this argument, it did not lead one very far in opposing GM food products.

Conclusion

Does vegetarianism preclude eating GM foods? More specifically, does adherence to vegetarianism affect the acceptance of plant-based GM food products containing transgenes from animal origin? Drawing from a delicate balance between theory and practice, I contend that genetic engineering does not constitute a direct threat to vegetarianism. As I explained, the implications of the case for vegetarianism find no hold over genetically modified plants with added DNA from animal origin. Rather, the menace is indirect and, as I suggested, comes from the potential underlying conflict between genetic engineering and vegetarians' system of values and beliefs about the social and physical environments.

Still, before reaching one's final judgment, another element must be contemplated. I underscored in the first section a distinction between "health vegetarians" and "ethical vegetarians." The distinction emphasizes the basis of vegetarians' primary reasons for espousing a plant-based diet. Most commonly, people who are prompted by health reasons to adopt a vegetarian diet progressively become aware of other justifications for doing so and strengthen their commitment to vegetarianism by moving from a single motivation to multiple motivations.[66] However, health vegetarians tend to be less committed ideologically to vegetarianism than ethical vegetarians.[67]

This means that their attachment to the belief that eating meat is morally wrong may not be very strong. It follows that "[p]eople who become vegetarians for health reasons are less likely to adhere to a strict definition of 'vegetarian' in practice, and it may be hard to inspire health-motivated vegetarians to commit to vegetarianism as the meat industry develops less-fatty meat products that consumers perceive as healthful and nutritious."[68]

Applied to genetically modified plant-based food products, the same line of reasoning suggests that health vegetarians are likely to oppose less resistance to the consumption of GM foods—whether they include transgenes from animal origin or not—if the safety of these products is clearly established and the latter offer proven health benefits in addition.

Considering the fact that health vegetarians constitute the majority of people who adopt a vegetarian diet, this weakens considerably the threat posed to vegetarianism by genetically modified plants containing added DNA from animal sources. Nevertheless, for those who identify strongly with vegetarian ideology, the menace is real.

Notes

The author wishes to thank Genome Canada and Genome Quebec for their financial support.

1. Claude Fischler, "Le comestible et l'animalité," in Boris Cyrulnik, ed., *Si les lions pouvaient parler—Essais sur la condition animale* (Paris: Gallimard, 1998), pp. 952–54.

2. Colette Méchin, "La symbolique de la viande," in Monique Paillat, ed., *Le mangeur et l'animal: Mutations de l'élevage et de la consommation* (Paris: Éditions Autrement, 1997), pp. 121–22.

3. Fischler, "Le comestible et l'animalité," p. 951.

4. Andrew Smart, "Adrift in the Mainstream: Challenges Facing the UK Vegetarian Movement," *British Food Journal* 106, no. 2 (2004), p. 80. See also Joan Sabaté, Rosemary A. Ratzin-Turner and Jack E. Brown, "Vegetarian Diets: Descriptions and Trends," in Joan Sabaté, ed., *Vegetarian Nutrition* (London: CRC, 2001), pp. 5–6.

5. Sabaté, ed., *Vegetarian Nutrition*, pp. 5–6.

6. Smart, "Adrift in the Mainstream," p. 80.

7. Donna Maurer, *Vegetarianism: Movement or Moment?* (Philadelphia: Temple University Press, 2002), pp. 15–18.

8. Ibid., p. 2.

9. The two focus groups were part of a study led by the University of Victoria Centre for Studies in Religion and Society. The purpose of the study was twofold. First, it aimed to develop a better understanding of the food and dietary practices of various religious, philosophical, and cultural traditions. Second, it aimed to determine whether those practices are affected by foods that have been genetically modified. Participants to the focus groups on vegetarianism were recruited through Earth Save Canada's e-newsletter, reaching more than one hundred vegetarians in the Lower Mainland region of British Columbia. Nine people attended the first focus group, and, out of these, seven attended the follow-up session. The author wishes to express her sincere thanks to all of them. The discussions that took place were extremely rich and informative and have helped her greatly in writing this chapter. Because of limitations in space, she regrets not having been able to incorporate to a fuller extent the wealth of ideas that were articulated by focus group participants.

10. Marjaana Lindeman and Minna Sirelius, "Food Choice Ideologies: The Modern Manifestations of Normative and Humanist Views of the World," *Appetite* 37 (2001), p. 176.

11. Jean-Pierre Poulain, "Mutations et modes alimentaires," in Monique Paillat, ed., *Le mangeur et l'animal: Mutations de l'élevage et de la consommation* (Paris: Éditions Autrement, 1997), p. 119.

12. Jean-Pierre Poulain, "Ces aliments bannis ou mal aimés," *Sciences humaines* 135 (2003), p. 38.

13. Lindeman and Sirelius, "Food Choice Ideologies," p. 175.

14. Colette Méchin, "La symbolique de la viande," in Monique Paillat, ed., *Le mangeur et l'animal: Mutations de l'élevage et de la consommation* (Paris: Éditions Autrement, 1997), p. 130.

15. Maurer, *Vegetarianism?* p. 2.

16. Lindeman and Sirelius, "Food Choice Ideologies," p. 182.

17. Laurence Ossipow, "Aliments morts, animaux vivants," in Claude Fischler, ed., *Manger magique: Aliments sorciers, croyances comestibles* (Paris: Éditions Autrement, 1994), pp. 127, 135.

18. Maurer, *Vegetarianism?* p. 71.

19. Linda Kalof et al., "Social Psychological and Structural Influences on Vegetarian Beliefs," *Rural Sociology* 64, no. 3 (1999), p. 501; Sabaté, Ratzin-Turner, and Brown, "Vegetarian Diets: Descriptions and Trends," p. 8; Smart, "Adrift in the Mainstream," p. 81.

20. Neal Barnard and Kristine Kieswer, "Vegetarianism: The Healthy Alternative," in Steve F. Sapontzis, ed., *Food for Thought: The Debate over Eating Meat* (Amherst, NY: Prometheus Books, 2004), p. 47.

21. Ibid., pp. 47ff.

22. See Jennifer Jabs et al., "Model of the Process of Adopting Vegetarian Diets: Health Vegetarians and Ethical Vegetarians," *Journal of Nutrition Education* 30, no. 4 (1998), p. 196. These nutritionists first introduced the distinction between "health vegetarians" and "ethical vegetarians" as a label to emphasize primary motives in the adoption of a vegetarian diet.

23. Kerry S. Walters and Lisa Portmess, ed., *Ethical Vegetarianism: From Pythagoras to Peter Singer* (Albany: State University of New York Press, 1999).

24. Ethical vegetarianism is distinct from vegetarian dietary practices inspired by religious doctrines such as Hinduism, Buddhism or Jainism. For present purposes, my interest rests exclusively with ethical vegetarianism. Religious teachings about abstention from meat are discussed in other chapters of this book. See also Steve F. Sapontzis, ed., *Food for Thought: The Debate over Eating Meat* (Amherst, NY: Prometheus Books, 2004), pp. 168ff.

25. See Walters and Portmess, *Ethical Vegetarianism*, pp. 11–12; Colin Spencer, *The Heretic's Feast: A History of Vegetarianism* (London: Fourth Estate, 1993), pp. 33ff.

26. Spencer, *The Heretic's Feast*, p. 43.

27. Ibid.

28. Walters and Portmess, *Ethical Vegetarianism*, p. 11.

29. See Sapontzis, *Food for Thought*, pp. 70ff.

30. Maurer, *Vegetarianism?* p. 76.

31. Frances Moore Lappé, *Diet for a Small Planet* (New York: Ballantine, 1971), p. 8 (emphasis in original).

32. Jon Wynne-Tyson, *Food for a Future: The Complete Case for Vegetarianism* (New York: Universe Books, 1979), pp. 16–17.

33. See Carol Adams, *The Sexual Politics of Meat: A Feminist-Vegetarian Critical Theory* (New York: Continuum, 1990); Carol Adams, *Neither Man nor Beast: Feminism and the Defense of Animals* (New York: Continuum, 1994); Carol Adams, *The Pornography of Meat* (New York: Continuum, 2003). See also Sapontzis, *Food for Thought*, pp. 247ff.

34. Jabs et al., "Health Vegetarians and Ethical Vegetarians," p. 196; Kalof et al., "Social Psychological and Structural Influences on Vegetarian Beliefs," p. 501; Poulain, "Ces aliments bannis ou mal aimés," p. 40; Smart, "Adrift in the Mainstream," p. 81.

35. Vegan/Vegetarian Focus Group, January 12, 2005, Vancouver, BC.

36. Ibid.

37. Ibid.

38. Ibid.

39. Ibid.

40. Vegan/Vegetarian Follow-Up Group, March 3, 2005, Vancouver, BC.

41. Empirical research would be required in order to establish whether vegetarians commonly hold the argument from the disconnection between consumers and food sources. Were that to prove the case, then this would demonstrate the interconnectedness between the vegetarian ideology and the contemporary critique of food production as whole. Vegetarians would thus constitute a particular group of consumers who are displaying toward foods the feelings of anxiety, insecurity, and defiance to which leads, according to sociologist Claude Fischler, the belief in a perceived distance between consumers and food sources. Claude Fischler, "Le consommateur partagé," in Monique Paillat, ed., *Le mangeur et l'animal: Mutations de l'élevage et de la consommation* (Paris: Éditions Autrement, 1997), p. 137.

42. Ibid., pp. 136–37.

43. Jean-Pierre Poulain, "Mutations et modes alimentaires," pp. 104–06, 113–14.

44. Arthur Caplan, cited par Gregory Lamb, "GlowFish Zoom to Market," *Christian Science Monitor* 22 (2004). Retrieved on September 27 2005 from www.csmonitor.com/2004/0122/p14s02–sten.html.

45. Gary L. Francione, *Animals, Property, and the Law* (Philadelphia: Temple University Press, 1995), p. 172.

46. According to philosopher David DeGrazia, a being has moral status *if and only if* that being's interests have some moral weight independently of the way in which protecting those interests redounds to the benefit of others. "Equal Consideration and Unequal Moral Status," *Southern Journal of Philosophy* 31 (1993), p. 25. Otherwise, although it might be recognized that this being's interests matter altogether, the consideration afforded to them only follows from the pursuit of an ulterior, beneficial end. DeGrazia defines an "interest" as "something that figures favorably in the welfare, good or prudential value profile . . . of a particular individual." Ibid., pp. 17–18. Examples of interests arguably shared by humans and animals include the avoidance of suffering, freedom, and life. Ibid. p. 18.

47. Vegan/Vegetarian Follow-Up Group, March 3, 2005, Vancouver, BC.

48. /Ibid.

49. See Tom Regan, *The Case for Animal Rights* (Berkeley: University of California Press, 1983), pp. 205–06, 208–10, 233–36, 247; and L.W. Sumner, "Animal Welfare and Animal Rights," *Journal of Medicine and Philosophy* 13 (1988), p. 159. See also DeGrazia, "Equal Consideration and Unequal Moral Status," p. 25.

50. Vegan/Vegetarian Follow-Up Group, March 3, 2005, Vancouver, BC.

51. Ibid.

52. Vegan/Vegetarian Focus Group, January 12, 2205, Vancouver, BC.

53. Vegan/Vegetarian Follow-Up Group, March 3, 2005, Vancouver, BC.

54. Ruth Harrison, *Animal Machines: The New Factory Farming Industry* (London: Stuart Vincent, 1964).

55. See for instance Peter Singer, *Animal Liberation*, rev. ed. (New York: Avon Books, 1990), pp. 95ff.

56. Maurer, *Vegetarianism?* pp. 58–60.

57. See ibid., pp. 58–62.

58. Vegan/Vegetarian Follow-Up Group, March 3, 2005, Vancouver, BC.

59. Ibid.

60. Vegan/Vegetarian Focus Group, January 12, 2005, Vancouver, BC.

61. Vegan/Vegetarian Follow-Up Group, March 3, 2005, Vancouver, BC.

62. Vegan/Vegetarian Focus Group, January 12, 2005, Vancouver, BC.

63. Vegan/Vegetarian Follow-Up Group, March 3, 2005, Vancouver, BC.

64. Vegan/Vegetarian Focus Group, January 12, 2005, Vancouver, BC.

65. Vegan/Vegetarian Follow-Up Group, March 3, 2005, Vancouver, BC.

66. Maurer, *Vegetarianism?* pp. 4–5; See also Jabs et al., "Health Vegetarians and Ethical Vegetarians," p. 196.

67. Maurer, *Vegetarianism?* pp. 20–21, 125. One potential explanation may be the following: "The North American vegetarian movement has always promoted diet as a means of bringing about self-improvement rather than as a means of achieving a public moral good, and popular discourse on health and fitness has perhaps strengthened this emphasis on contemporary vegetarianism as a personal, self-benefiting choice." Ibid., p. 24.

68. Ibid., pp. 125–26.

4

"When You Plow the Field, Your Torah Is with You"

Genetic Modification and GM Food in the Jewish Tradition(s)

Laurie Zoloth

Introduction to the Problem: Two Cases at Hand

Professor Daphne Preuss is a thoughtful woman, used to watching small shifts in simple systems. She is a plant geneticist, and she wants to understand how genes transfer traits and maintain species identity using complex triggers and controls so that consistent genetic cassettes of information and instructions for the proteins that create species-specific traits are precisely copied and inherited. As species mutate and evolve into new species forms, to maintain species identity, most organisms exercise elaborate controls to ensure that their offspring inherit the correct genetic material. When new species emerge, the machinery that regulates inheritance diverges rapidly, creating barriers that prevent cross-species hybridization. Her discoveries have highlighted how rapidly this evolution can take place and the puzzles about what keeps new plant species both stable and still able to adapt quickly.[1]

But Dr. Preuss has a different sort of question—an ethical one. She is interested in seeing if a genetic set of instructions can be inserted into the DNA of plants used for food in the world's poorest countries to make them "more useful." She is working to make muskmelons that carry genes to make proteins, and she tells me about work done by her colleagues in a project called "BioCassava Plus." In this project, a multidisciplinary team of international researchers in science, sociology, and anthropology is working to both create a nutritious crop for sub-Saharan Africa and to anticipate the impact of a genetically modified food in that culture. Inadequate nutrition is the single greatest cause of the steadily rising mortality, morbidity, and suffering in sub-Saharan Africa, which is facing a rapidly changing climate. Two hundred and fifty million Africans rely on the starchy root crop cassava as their staple food—which is itself an irony. Cassava roots are not native to Africa. They were brought there by missionaries from South America in an early effort to provide food. But they were not a fortuitous choice, for while they are easy to grow, they have the lowest protein-to-energy ratio of all the staple crops, with a typical cassava-based diet providing less than 30 percent of the minimum daily requirement for protein and only 10 to 20 percent of the required amounts of iron, zinc, vitamin A, and vitamin E, and they have high levels of harmful sugars. The goal is to genetically alter the cassava so that people who can afford to grow only this one crop would get everything they need nutritionally, without the harmful effects. That will not be all: researchers are working with African traditional farmers to make sure the crop actually tastes good and is still easy to grow.[2] Musk-melons and cassava are interesting choices: they are nearly entirely in the control of women in villages. Where starvation is a constant threat, making these crops into a high protein, highly useful food would be of inestimable value. We are talking about genetically modified (GM) food, and Dr. Preuss wonders aloud about the acceptability of genetic changes that would make plants more resistant to drought or able to grow in salty soil.

It is an important question and a pressing one: August 2005 marked the final mapping of the rice genome, a foodstuff that has provided 20 percent of the world's human nutrition over thousands of years. Should rice be altered, enhanced, grown differently? Would this constitute a considerable change in the natural order of things, or is the genetic manipulation of vegetables, grain, and animals merely an extension of human primate behavior, a sophisticated version of the most basic command of God in the Hebrew scripture—to fill the earth and subdue it? It is a complex question, but one that, in the face of continuing conditions of famine and food shortages throughout the world, requires our thoughtful consideration.

The Second Case: The Vancouver Focus Groups

Far away from Pruess's Chicago lab, two groups of Jewish citizens gather to reflect on the issue of GM food. For the Vancouver Jews who partici-pated in the University of Victoria focus groups, one from Orthodox and one from each of the three non-Orthodox traditions (Conservative, Reform, and Reconstructionist), the discussion would be about the way in which lay knowledge is both constructed by and constructs an ongoing debate about genetic policy. These Jews, lay members of the community, are the public witnesses to the debate within this book. Here, in these groups, the issue of genetic modification of plants and animals was understood through a complex cultural, ethnic, and political lens—a mixture of memory, liberal political theory, and serious concerns about the future. In both groups, GM foods were viewed as part of a larger trend about which they were worried. Their concerns expressed a general sense of loss of control, a distrust of a science linked to the marketplace, and worry over how these how these developments affect them, perhaps without their knowledge. The religious movement, affiliation, and level of observance of the members of the Jewish group were nominally different: they identified as religious and nonreligious, as traditional or not. They all had expressed concern or interest in genetically modified food, and none of them was aware of positions taken by Jewish scholars about the topic. That the focus group was diverse in its practice was important, for how carefully one observes the biblical and rabbinic rules about food is a clear mark of distinction between Orthodox and some Conservative Jews who are committed to a robust set of rules that define what food is permissible (a system called "kashrut" or kosher), and others from movements within Judaism who believe the rules to be a more general guide to principles of behavior. They had much in common. They were all recruited from synagogues in Victoria or Vancouver, British Columbia, in Western Canada, and were all asked the following questions: Why are you here? What, if any, food and dietary practices are important to you in your Jewish faith?[3] The groups were then given a presentation on the basics about GM foods and asked what they thought of transgenic foods—meaning, in the cases described to them, those in which genetic instructions for making a particular protein to generate a particular trait are transferred from one living entity to another. In plants, such technologies are increasingly common; in animal products the cases were more hypothetical. The participants were then asked whether such practices would be permissible under Jewish law.

These Jewish citizens reacted as many North Americans do to GM food. They had, largely unwittingly, consumed it for years. Indeed, nearly every product that uses corn products, wheat, soybeans, or enzyme additives is based on GM technology, which includes 60 to 70 percent of all grocery store items.[4] Yet when asked directly about the use of GM food, they expressed three sets of fears: first about safety, second about its permissibility under the rules of kashrut, and third, whether it was a hubristic attempt to alter nature in radical ways, resulting in unknown outcomes. For these individuals, whose background and beliefs emerged from radically different locations within Jewish thought, the issues were complexly linked to how they understood both the process of genetic alteration itself to work and what they anticipated as the telos of the project. Interestingly enough, the rules of the system of food restrictions (kashrut) that governs food customs seemed to shape the response to GM concerns in both groups, despite the fact that few of the group kept kosher in any way. Yet such is the power of the rabbinic legal system that all of the participants understood why the first problem for Jews would be about the laws of consumption and production. "Would it be fair," they asked, "to change food in ways that might alter its status as kosher? What is the role and meaning of 'species'? What is a gene in any case? Could anyone be trusted to monitor GM food?"

All of the participants agreed that the best regulatory system was not the companies, for many in the groups thought them greedy and duplicitous, and perhaps not even the USDA, who some thought would deceive them, but some larger, international body. The level of mistrust of government, science, and the market was significant in this group of Canadians:

> I mean for me it's all about profit. It's money-driven completely. It's a way to make some money. It's for corporations to make money, and you know, they are not telling people the whole truth and we don't know what's in it and I know that from what I've been reading that they already have done controlled studies where there were certain things that they created that wiped out an entire field of crops forever so nothing would ever grow on that land again. And I'm thinking what if they had introduced that into the world, maybe it would have wiped out every plant on the earth. So I mean, this is pretty dangerous stuff that they are messing with. And to me, it's all about greed.

For these Jewish communities, all of whom are living at some remove from the large centers of Jewish population and scholarship, issues of halachah (Jewish religious laws and customs) are understood and inflected via the lens of modernity. This included participants' views about the labor practices, the genetic processing, and the science itself involved in genetic modification. Strikingly, the carefully articulated views in the scholarly literature about what the official rabbinic rulings relative to GM food are were nearly entirely unknown to these groups, and even more strikingly, many expressed an emphatically stated "folk" understanding of the reasons for kosher laws that were relatively removed from the academic sources. For example, many carefully explained the food restrictions as existing primarily for health or cleanliness reasons (e.g., that pigs are prone to disease, and milk-meat combinations are harmful), a claim that is not supported factually. (This was a view that was popular in the writings of nineteenth-century German Jewish scholars but largely dismissed as an attempt to try to justify and rationalize customs seen as primitive and absurd to Western Europeans. As such, it has widespread appeal). In actuality, the laws of kashrut are in fact complexly random and are carefully delineated in the tradition as a genre of law that *is by its nature* not justified or rationally based. This problem of the irrationality of the rules has been a deeply contested puzzle since rabbinic authorities debated the issue. And interestingly enough, in some cases, this folk view of kosher laws was overly (and incorrectly) restrictive: "From what I know, we're not supposed to have plants grown with other things such as you can't mix fish with plants. And you can't grow—I know that you are not supposed to grow apples with oranges. . . . I guess that is where the phrase comes from."

At times, the popular view was oddly (and incorrectly) lenient. Cheese is rendered kosher when it has a certification that it has not used animal rennin in its production. Yet when asked about what renders cheese kosher or not, the group agreed on a completely new idea:

(Cheese is not kosher if) the gelatin is nonkosher. Many times it's animal based. And unless it comes from a ritually slaughtered animal which is acceptable as kosher, then the cheese itself is not kosher.

(Moderator) So if there was gelatin from an animal that was kosher, that gelatin additive would be okay?

Yes.

Everyone in the group spent considerable time thinking about food and health and showed a lot of concern about its providence. The focus groups were drawn from Jews who already cared about food and thought it important to maintain rules and limits in its consumption, Jews who were interested in health and longevity, and Jews who were interested in issues of social justice or ecology. They had a variety of reasons for joining the group. One claimed to be "interested in, you know, potential effects or non-effects of GM foods and I thought that it would be an interesting topic to discuss." Others were already wary, stating "I've had an interest for several years. I've done a lot of reading on the subject so I am quite interested in it, in genetically modified foods, the implications, etc.," or, "I'm quite concerned about what GM foods, what impact if any do they have on our health." And, "I'm worried about GM foods because I don't know how it affects our health. Also I'm worried about kosher." One came (honestly enough), for the money offered. In a typical statement, one noted,

I'm worried about GM foods because I don't know how it affects our health. Also I'm worried about kosher and about people who distribute this food.

I have read several articles on some of the organisms that are used in modifying foods and most of them that I've read about are definitely nonkosher. They come from ocean organisms which are related to lobsters and crabs and types of shellfish and therefore would not be kosher, and yet these foods are not labeled. That puts a lot of us at a disadvantage.

Food in general was seen by all the groups as central to Jewish life and culture. This was expressed in two ways. First, it was expressed for nostalgic reasons, recalling the yearning for an immigrant past or a time of greater certainty or innocence, where food was harvested and cooked by mothers at home. They also understood the religious holidays and rituals of Jewish culture as occurring in the context of large communal gatherings, all of which needed food to be complete. There was a repeated sense that the rules of kashrut had an independent worth as a heuristic for specific values: kindness to animals, respect for the environment, and limitations on unbridled consumption. Yet there were countervailing comments that expressed ambivalence about the regulation of science. Some understood the long tradition of science and medicine within Jewish intellectual life, with one participant wondering aloud if "too much fear or regulation would have hampered Einstein" and with others noting the strong value on healing and restoring the world within the tradition and the way that genetically enhanced food might play a role in such restoration:

> We have a roadmap, we have a guide, the Torah, which is given to us by God, and it shows us the broad parameters, what is the framework within which we are free to operate. And there are—there are certain guidelines. There are certain red lines, if you like. These are the things you must not do. There are cases where it's a matter of saving life. Most of the rules are set aside for that, right? But we are talking about here now—you know—we all have full bellies. We live in a society, thank God, that none of us are starving or need to starve. We can make free choices, and what does our tradition teach us rooted in the Torah? How far are we able to go, and—until here. And over there, we can't. As convenient as it might be, as logical in the short term as it may be to everyone, no, our tradition says there are limits.

Curiously, not one of the participants raised a core issue that the rules of kashrut limit what Jews can eat or breed, not what the rest of society can eat or breed. None raised the fact that Jews are permitted to use animals and eat fruit that have been bred by gentiles in impermissible ways. It is proper to ride a mule or eat a nectarine, for example.

But perhaps most striking is the fact that none of the participants was aware that GM foods and the issue of artificiality in food production are largely settled problems for the leadership of the Jewish community, including the oversight committees that determine the international standard of kashrut itself. Consider this comment by Mark Washofsky, reprinted in the *Chicago Jewish News*:

> According to most rabbinic scholars, genetic engineering is not a violation of Jewish law. It is not considered a transgression against the natural or the divine order of things because the Torah consigns the natural world and the human body as a part of that world to human control . . . Moshe ben Nachman notes this license includes the power 'to uproot and to destroy' and to otherwise manipulate the created universe for our own benefit. . . . True, there are some prohibitions against the mating of 'diverse kinds' (kilayim) of plants and animals . . . however the leading Israeli halachist, Shlomo Zalman Auerbach, taught that biblical verses refer specifically to the mating of animals and to the mixed seeding and grafting of plants, not to modern technologies of genetic engineering, and his views have become predominant in the subsequent halachic literature.[5]

Consider also this, from Avram Riesner: "Use of the products of hybridization is affirmatively permitted. In fact, many hybrids are presently on the market, both hybrids of different strains of the same kind of plant, which would not be kilayim, and those of separate species, which would be considered kilayim, the product of agricultural and animal husbandry techniques honed before the advent of genetic engineering, No product is banned. Indeed, this is not even a modern leniency, having its earliest source in the Tosefta."[6]

To be sure, there are some who are uneasy with genetic engineering, viewing it as a part of a long list of modern interventions that are seen as part of a problematic crisis. As Arthur Waskow writes,

> And the world-wide ecological crisis created by the actions of the human race in the last few generations of modernity has begun

to raise concerns among Jews as well as other communities, for
how to redefine what food is proper and sacred to grow and to
eat. Questions about the use of pesticides and of genomic engineering; of the burning of fossil fuels to transport foods across
the planet, meanwhile disturbing the whole climatic context in
which the foods are grown; the misuse of topsoil and the use of
long-term poisonous fertilizers; the effects of massive livestock
breeding on production of a potent global-scorching gas, meth-
ane—all these have raised profound new questions.[7]

While the ideas behind this quote from Arthur Waskow were expressed
by many, most reported learning about them from the popular press.

How are we to evaluate such conversations when the logical discussants—scientist, community, and religious leadership—seem to be in different rooms entirely? It is not a new problem, for it is the premise of
rabbinical law that questions emerge from actual dilemmas in practice. The
voice of the community has long played a role in the debate. So let us turn
from the concerns of the focus groups to what might be the source of a
useful response.

Facticities of Talmudic Arguments

We are led to ask, then, two questions: on what basis do the rabbinic
authorities who determine kashrut derive these rulings, and why does this
selection of Canadian Jews find them counterintuitive? It is not, of course,
only this focus group. Waskow and others have linked GMOs to a network
of concerns about modernity and the global market. Jews, often identified as
such, are participants in protests against GMOs. The first reviewer of this
article, and the participants in the public outreach forum that accompanied
this research, raised the same fears about GM food, profits, and contagion.
Yet it is my contention that the texts support very different concerns, ones
that emerge from the deepest sources and oldest arguments about food,
justice, and the sacred approach to God within the tradition.

In reflecting on how the tradition of Jewish law understands the question of GM food, I began these two practical conversations, one with a
scientist, and one with a concerned community, for a distinctively Jewish
reason: the method, goal, and meaning of Jewish ethics is rooted in practical questions that are raised within living communities that exist within
the imperfect, chancy, and often isolated conditions of exile. The traditions
of Jewish thought are not concerned with abstract norms, nor what might
yet be the case, but with how moral dilemmas present actually—in specific

moral locations, with specific moral features, and with attention to specific and competing moral appeals. The focus groups were asked questions about theoretical problems, ones that had not yet come to pass and may indeed be impossible, for example modifying a pig until it is an animal that meets the requirements of kosher rules—which would be one with split hooves that chews its cud. A very different conversation is possible if the group were presented with both genetic problems, for example, square tomatoes, and genetic successes, for example, crops that allowed farmers to use far less pesticide or the work unfolding in Dr. Pruess's lab. Yet despite the uncertainty about the use of genetic modification, one core value did in fact emerge quite clearly. All of the participants were clear that genetic modifications that were lifesaving would be permissible, even laudatory. Why is this the case?

Why does turning to the facticity of the need of the Other interrupt every abstract discourse, and why, ultimately, does raising the needs of the Other against the fears of the self so fundamentally construct the debate? I will argue in this chapter that it is both this exquisite attention to the needs of the other and a careful analysis of the actual science of gene transfer that grounds Jewish views on GM food.

The complexities of the texts within Jewish law allow a rich and multivocal response to a modern concern: *May we make the world?*

Facticities of Molecular Biology: What Does "GM" Mean?

The concerns about genetic modification need to be taken into account against important scientific realities of the process. The scientific details are critical in the academic and scholarly discourse that informs halachic decisions, and from rabbinic times, the technical aspect of the matter is intended to be the start of debate. This was true in Maimonedes, who used Aristotelian and Arabic science, in the later commentators, who relied on post-Enlightenment discoveries, and in the contemporary era, when the kasrut authorities use modern genetics.

It is the current and widely agreed upon theory that all living organisms, from yeast to humans, are organized in the same manner: all are made of specialized and differentiated cells (from one to many millions) with a cytoplasm, bodies within the cytoplasm, and a nucleus, which contains within it a construct of double-sided molecules, DNA, which is a chemical assemblage of four acids (ACTG) spiraled around a sugar linkage ladder. The molecules of DNA can be copied to make duplicate strands, and hence, when the cell grows then divides, the DNA is identical in both nuclei. DNA contains the potential to assemble sets of proteins, and these proteins drive, by their combination and confirmations, the production of enzymatic chemical processes that allow the cells to function—to take in nutrients, to make more cells, to move about, to reproduce, to signal. Most of the cellular functions are similar in most organisms, and speaking in an evolutionary manner, once a problem (such as how to metabolize sugar) is successfully solved in one organism, the genetic solution is largely passed on to other organisms that emerge along the phylogenetic lineage, which is why we share so many genes with the tomato plant or fly in the first place. Genetic characteristics define the ontology of the organism. It was the insight of Darwin that the species we see so clearly around us—celery, oak, and orchid—are the result of chance variations in type, a changed ecologic niche, and the reality of abundant reproduction in a world of limits. Desirable traits are passed along to offspring because the organisms that bear them have (or are given, in the case of agriculture) the best chance of successful reproduction and heritability. Successful variations are retained, and when characteristics are altered thoroughly and permanently they become different "species" (a term for organisms that do not interbreed using normal sexual reproduction because they are so biologically different).

Since the late nineteenth century, the variations have been understood as carried by units of heredity called "genes" and since 1953, when Watson, Crick, and Franklin described the molecule DNA, as carried by segmented DNA on chromosomes. Now that the DNA of many creatures has been carefully mapped, it is clear that many entire segments of DNA are highly conserved across the lineages and across the species. DNA in plants, yeasts, fruit flies, worms, mice, and humans is made of the same base pairs—and is functionally identical—hence the idea that this is a "mouse" gene, and this a "fish" gene can be misleading. DNA in this way is like an alphabet with cognates in many languages.[8] Further, since the function of DNA is to make proteins, altering or inserting a small triggering molecular change can change the proteins made or allow the DNA to express in a different way—and this too is at the basis for how we understand evolutionary change.

Restriction enzymes allow scientists to cut and splice the DNA into small segments, add the sections to other strands of DNA, and allow the new patch or "cassette" to express the proteins it is capable of expressing, even in a different cell nucleus, usually a yeast or ecoli, which then transfers the gene sequence into the germ line of the developing embryo. It is in part this process of "copying and recopying in various media" that allows for the ruling of the acceptability of the process.[9] Additionally, it has been long accepted that when a salubrious mutation happens naturally, the plant or animal can be deliberately bred to retain the trait. The mechanism that drives those spotted sheep into Jacob's flock may be the "peeled stick" of the biblical text, but it is carried out as a mutation in the melanin gene.

Corn cultivation—all of which was done manually—involves the placing of the DNA of one plant on the other, hoping it will mix correctly. Hence, all breeding is genetic manipulation. DNA manipulation could be understood in this manner as largely a close and very precise view of the method of all selective breeding.[10]

Thus, keeping the specificities of the science in mind, let us turn to the system of Jewish ethics.

A Brief and General Introduction to Jewish Ethics

The Jewish tradition is a long complexity, and in that complex tradition, there are several themes and inflections on the topic of food. For the reader unfamiliar with Jewish thought, however, a brief methodological compass is needed. The Jewish system of ethics is based on a two-thousand-year conversation that interrogates the texts of the Hebrew scriptures and applies the interpretations and the arguments of interpretation to contemporary and specific quotidian dilemmas. There is no central personage or magisterium in control—the texts themselves (Torah or Hebrew Scripture, the Talmud) play that role—but there is a history and lineage of teachers and decisors, including Maimonides, Rashi (in the medieval period), and Joseph Caro (in early modernity). In the modern era, Judaism divides into four different denominations: Reform or Liberal, Conservative or Masrati, Orthodox, and Reconstructionist/Renewal, movements with differing teachers and commentators. In all cases, the casuistic core of ethics is maintained—with recourse to proof texts as the primary rhetorical move in all argument. The interpretations then translate to praxis: kashrut is determined by an actual rabbinic authority (in the modern period, one trained in chemistry and agricultural technology), who goes to every setting in which animals are killed; food is processed; or food is packaged, shipped, stored, or served. This insures an ongoing, inspected kosher status.[11]

For the Orthodox Jews, the most widely accepted symbol of kosher status is the OU or Orthodox Union's own *hechser* (mark). The OU maintains a large staff of rabbis who fly about internationally, checking on thousands of processes and additives, including genetically engineered ones, providing something precisely akin to the independent, international oversight that the Jews in the focus group were seeking. Genetic modification has followed the same methodological path as all new technology, from the bicycle to nuclear energy. Where the legal decision, called the *halachahic* decision, about the technology's permissibility emerges from a widely understood and clear precedent case, the commentators will have to decide how much weight it carries and how firmly to adhere to the biblically or rabbinically commanded law as it has been interpreted across medieval and modern times. In the case where the legal texts are silent—as they are in many cases of new technology—the interpretation of contemporary authoritative scholars is sought and then widely distributed. Then scholars translate such texts, and pulpit rabbis learn to teach them. What role should general admonitions play? Should the idea that if a thing is not mentioned as forbidden then it is permitted be applied? For GM food, the intricate science, general philosophical admonitions about the care for the earth and the care for animals, and the general standard that everything that is not expressly prohibited by the text is permitted have all been considered.

Process and Deliberation

For many commentators, the goal of the commandments given by God in the scriptural texts is complex. Argues R. Saul Berman:

> Laws are given to remind us of our existential moral location. We are in/and of a world of nature created by God, yet peculiarly and clearly imperfect, unfinished, and mutable. We are expected to be alert to social justice issues, to construct a social and moral economy, for we live in community, and we are expected to learn to live within limits and constraints. We are told that the most ordinary of acts occurs both as utterly observed by God and witnessed by the community—hence, ethics, and hence keeping the commands of kashrut, is both a theological and "published act"—like liturgy, it is done in community. As Waskow and others have noted, food offering and practices were a core exhibition of the public "nearing" to God.[12]

Ethical norms are not arrived upon by a process of reflection on principles or on rational interpretations of natural events as in many other systems, such as reflective meditation or natural law theory but are considered as part of a very concrete, detailed, and systematic law given at Mount Sinai to the Hebrew slaves directly after their liberation. The subject of the biblical law, including the details of the puzzling rules about consumption, has been the subject of generations of interpretation and commentary, from the rabbinic period during which the Talmud was composed until the present day, in which living communities need to make decisions about emerging scientific and technological choices. For the purposes of this review, it is important to note that the laws vary in an important and obvious regard: some have seemed to generations of commentators and readers of the contemporary period to be utterly logical—universally derived from the human condition (put up guard rails on high places, wash after you touch a dead person, do not plot murder), while some seem bafflingly obscure or arbitrary (do not consume milk and meat together, but you can drink milk with fish; do not wear clothes that mix wool and flax, but you can wear clothes with cotton and flax). Such laws are called "hukkim"—or laws that are given without rational explanation. Many of the laws that concern food and the alteration of food are such as these (although in the nineteenth century, rationalists attempted to understand these as hygienic or anti-infective regulations), and in the contemporary period, this effort has included new ideas about justice, or organic food, or vegetarianism). Notes Waskow:

> The new technologies of modern life have thrust upon the Jewish community still another unexpected question: Should the category 'kosher' be reserved for food alone, for what we literally put into our mouths and gullets?—Or would it make sense in our generation to apply the basic concept to other products of the earth, other forms of 'eating'—consuming? If the whole notion of 'kosher' food emerged when most human beings were farmers and herdsfolk, eating food, should there be a new kind of 'kosher' for a world in which human beings also 'eat'/consume coal, oil, uranium? Should we attempt to imagine an 'eco-kosher' code for consuming not only food but also all that an abundant and partially depleted earth produces?
>
> Through the peculiar history and theology in which the Jews preserve both a sense of indigenous earth-connected peoplehood and a sense of worldwide presence and significance, can the Jewish attention to food as a crucial means of connection with God play an unusually useful role in the future of the human race and planet earth?[13]

Rabbi Saul Berman notes that the application and study of these laws has long been a careful practice, in part *as* a practice of discipline and rule-following. He suggests, however, that we look for more than merely disciplinary training in following these laws. Berman suggests at least two purposes in the laws that we will be closely examining—the laws of agriculture and cultivation, first, to remind us that God has created the natural world and that God's presence ought to be seen in all the work of the land, and second, that maintaining agricultural order is also maintaining a just social order.[14]

Moreover, the system of the law is precise. This allows for two preconditions for ethical response. First, while Jewish law in general takes a cautious initial stance toward innovations in the scientific or social realm, when it regards a specific case, the halachah follows a delineated and narrow path, for example, understanding the categories of animals noted to be *trief*,

or forbidden, as limited to the exact ones in the biblical text. As Rabbi
Avram Steinberg notes, "[o]n a fundamental level. . . . anything that there is
no reason to forbid is permissible, and needs no justification. For the Torah
has not enumerated all permissible things, rather, forbidden ones" (Tiferet
Israel 4.3). For Steinberg and others, what is at stake in the consideration
of cases in Jewish ethics is whether an inherent and specific transgression of
halachah will occur in the actual process of the case or in its results.[15] As
Miryam Wahrman notes, the very oddity of the *hukkim* make their application more precise, which is important for our case:

> Of course, any procedure approximating DNA technology is not
> mentioned in the text. All laws in Mishnah Kilayim are con-
> sidered to be Hukkim or statutes for which God provides no
> explanation. According to tradition, Hukkim are derived direct-
> ly from the Bible and must be adhered to without modification.
> And since these Hukkim are so specific, and no explanation is
> provided for their enactment, rabbis insist that each huk be fol-
> lowed precisely as described, to the letter of the law. They may
> neither be added to nor subtracted from. Thus, although the
> Mishnah directs that the drafting of plants is prohibited, this
> does not necessarily prohibit the isolation of specific strands of
> DNA sequences and the chemical joining of those DNA strands
> to DNA from another species.[16]

Second, the law as given at Sinai is a law for the Jewish people, not
for all of humankind, which lives outside that particular covenant. Hence,
even if something is not permitted by the *hukkim* to Jews to make or to do,
it is not intrinsically immoral unless it violates the Noahide laws mentioned
as given to all peoples as Noah's family emerges from the ark. This is why,
for example, certain sorts of cross-breeding or grafting are prohibited for a
Jew, but a Jew may ride a mule, eat an apricot, or happily buy broccoli.

A third general overriding aspect to Jewish ethics and innovation
involves the core values of saving a life (*pekuach nefesh*) and healing the
essential brokenness of the world (*tikkun olam*). For Jewish tradition and
halachic response see the world as largely unfinished and in need of repair—
repair that is enacted by direct and concrete human interventions in illness,
suffering, agriculture, and industry. While an occasional argument for "natural law" and "natural order" is raised in the texts, in the work of medievalists
Maimonides and to a larger extent Nachmonides (Rambam and Ramban)
there is lasting argument against it by the majority of Jewish commentators
who make the claim that the role of a Jew is to be part of the ongoing act
of creation. Nowhere is this truer than in the idea that healing the ("naturally" occurring) illness and injuries is a core task of the Jew.[17]

Hence, let us note a tension in our task. First, since nothing about genetics, or indeed about the entire world of cellular or molecular biology, could have been described in the Talmud, it is perhaps the technical case that no restrictions could be applied to its use, and since, further, the halachic considerations only apply to what Jews can make, it is perhaps the case that Jews, while not able to be genetic engineers, can partake of any of the "fruits" of genetic engineering. Second, if there is a primary concern for healing interventions, perhaps anything that fits that criterion (which surely the provision of ample nutrition must) would gain a sympathetic hearing halachacly. These considerations have led to a broad and far-reaching acceptance of basic research in genetics; the acceptance of genetic modification; and advances in human, animal, and plant reproductive technology.

Yet the idea of altering nature and natural kinds has surely been discussed in the literature, and are we perhaps permitted some speculation on these narratives, even perhaps drawing on their powerful metaphors for some caution in the face of singularly robust or de nova research? Can we turn to general midrashic accounts to create responses in Jewish ethics, such as general admonitions not to harm nature or act in a hubristic fashion? Should our capacity for failure and error turn us toward texts of error, such as the Tower of Babel, as some suggest? Thus we see a fascinating paradox in Jewish responses to GM food. The most authoritative and conservative halachic authorities—including the majority of Conservative and virtually all of the Orthodox—advocate the widest use of new ideas and technologies, while the most liberal thinkers—those within the Reform tradition, the ones usually most likely to call for the most creative use of texts—advocate for the most cautions on the use of GM food.

A General Introduction to the Literature and Themes

Given this doubled voice of Jewish ethics, I will represent the range of possible questions and sources for response to our case, noting first the larger inflections of the literature and traditions that may refer to food use and alteration, then the classic halachic texts, and finally, I will explore one of these texts—that concerning mixed seeds—more closely, to see if a method of textual reasoning helps us to understand one answer to our case. The following are five contextualizing issues to note in understanding the place of food in Jewish life.

Desire and the Limits of Desire

The concerns with food and consumption begin in the Book of Genesis and the account of the forbidden fruit. In this story, food is a symbol for desire, for knowledge, and for transgression. Flung from the garden, humans must take up the task of food production and must pull the very fruit out of the ground with the labor of their bodies, linking food, work, and childbirth generatively in one text. In the Abramic story, the patriarch cooks for the divine visitors who tell him of Isaac's birth. Food in this account is hospitality. In Exodus, the story of Jacob also links the desire of Esau with hunger, and it is hunger that disrupts the basic order of primogeniture. Jacob manipulates the goats (genetically) to produce a herd with spots (a core source text for the permissibility for genetic engineering). In the Joseph story, grain, abundance, and famine are the operative symbols of dreams and prophecy. Joseph's family is then driven by hunger toward the paradox of redemption that will also begin bondage. Note that in all of the stories, nature is decidedly not normative: nature and production of food are instrumental. Finally, in Leviticus, the people yearn for food, desiring meat and leeks and onions, rather than manna. In this narrative, manna—odd, strange food that it is—is also a symbol for Torah, linking wisdom with sustenance.

Food as a Part of the Sacrificial System of Forgiveness, Joy, Justice, Order, and Piety

Food in Leviticus and Numbers shapes the entire sacrificial system. It is the way that communities and persons approach God: the word for sacrifice, *korban*, has a literal meaning of "coming near."[18] The premise of the Temple system is that sacrifice of animals (and grain and plants) allows for celebrations of holidays, forgiveness of error and transgression from the law, and observance of a daily pattern of interaction with the sacred. The Levites, who are not permitted to own their own land, serve in the Temple and eat the meat that is offered. (Not in all cases; some is entirely burned). Meat is not eaten casually but is always a part of a ritual activity in some measure. Passover is remembered with the act of participatory sacrifice and a long intricate meal, with all food having a meaning. On the Sabbath, the order of eating is fixed: two loaves of bread at three meals, which are dinner, lunch, and third meal. In contemporary times, holidays are marked by a *suedah*, or ritual meal.

Food and the Rules of Production Are the Essential Way That God Cares for the Poor via Our Labor

Rules of agriculture define a system in which the corners of the field, the gleaning from the harvest, are due to the poor as a straightforward part of what the poor are owed. All holidays are linked to the issue of what food to give to the poor. At the holiday of Purim "*mishloat mannot*" is exchanged, which specifies what food to give, in an enactment of the story and in a reestablishment of the moral economy of giving. Thus the exchanges of food, how food is produced and offered, and who is invited to the holiday meals in Jerusalem are the very substance in which the enactment of justice is described. Our labor is transformed into food, which is transformed into justice.

Rules of Kashrut as a Complex and Elaborated System of Eating

The system of kashrut (keeping kosher) moderates desire, production, and consumption and is a series of limits of what is "fit" and what is decent to eat.[19] The rules are complex and nonrational, sometimes linked to moral criteria (no prey animals are eaten) and sometimes not (split hooves and cuds). In fact, the intricate rules of kashrut are seen by some as the very sort of rule that, since they make no "common" sense, are to be followed in a disciplined manner to teach discipline itself.

Strong Cultural Norms for Food as Hospitality and Social Solidarity

The loss of the Temple and the sacrificial system removed many aspects of public worship from Jewish practice and moved the moral location in which the commandments were enacted to the home and, to a limited extent, the marketplace. Since modernity calls for a secular set of rules for marketplace interactions, it is the home and the private sphere in which liturgy and tradition are experienced by many in the Jewish community—and it is here where food customs most entirely impact. Hence, sharing food, eating as a part of festivals, and the ethnocultural aspects of food allow for food to be a rich carrier of symbolic meaning. The meaning is complex but shares core features with other immigrant communities: the sense of triumph over depravation and adversity, the idea of hospitality, the idea of community norms, and the idea that women's skills can be of aesthetic, social, and moral value. Of note is the fact that for centuries these food practices meant that Jews could not share a meal with non-Jews.

Review of Halachic Literature concerning Genetically Modified Animals, Grain, Vegetables, Fruit, and Dairy Products

The major legal concern of the halachic system concerning genetic engineering in general is that Jews are forbidden in the Hebrew scriptures from mixing (some) animals and (some) plants and even wearing a mixed garment.[20] Here are the key passages:

Leviticus 19:19

"You shall keep my Laws. You shall not let your cattle mate with a diverse kind; you shall not sow your field with mingled seed; neither shall a garment mingled of linen and wool come upon you."

Deuteronomy Chapter 22: 9–11

"You shall not sow your vineyard with different seeds; lest the fruit of your seed which you have sown, and the fruit of your vineyard, be defiled. You shall not plow with an ox and an ass together. You shall not wear a garment of different sorts, like wool and linen together." This is understood, generally as a prohibition on *kilayim* or forbidden mixtures. It is the subject of an entire tractate of the Mishneh, also called *Kilayim*. In this tractate, the kinds of mixings are intricately defined (which animals are not to be bred, which seeds are not to be mixed, and how to check for infractions in the laws of *shatnez* (combing, spinning, and weaving, which technically refer to the human activities that are needed in wool and linen production).

The premise of these passages is that creation has an intent, a view expressed in the thirteenth century by Nachmonides, who wrote that "the reason for *kilayim* is that God created species in the world of living creatures, the plants and the animals and . . . commanded them to reproduce their own kind and they shall not change as long as the world exists."[21]

What is at stake is the sense that it would be arrogant to assume God needed help in creating the world. Yet that is precisely the view of many who argue with Nachmonides—for most Jewish commentators clearly saw that there was variation, breeding, and active change in the argricultural world and that creation is clearly a work-in-progress. R. J. David Bliech and others (including earlier Talmudic rabbis) argue strongly for the sense of the world as unfinished, and acts of manipulation are merely continuing the Adamic project—subduing the world with the technology that God has shown to us.

Hence the question of the normativity of nature is a largely settled one in Jewish thought: nature is morally neutral. But what of the other major question: is genetic engineering considered forbidden mixing, *kilayim*? For most, it is not. "A leading Halachic authority in Israel, Rabbi Shlomo Auerbach, supported the use of genetic engineering; he did not believe it constituted *kilayim*. Dr. Avraham Steinberg, the current leading Halachic authority in medical ethics, reports Auerbach. Our view is that genetic material may be transferred from one species to another by means of bacteria or viruses and the resulting species not considered forbidden mixture."[22]

Many have noted that not *all* mixtures are prohibited, but only the ones carefully noted in the Gemora. Further, size matters in the halachah. "A minority becomes annulled in a majority, or a major disannuls a minor quantity or the lesser is canceled by the larger," is noted as important by a number of commentators on this problem. Since many plants are already hybridized (broccoli), and since this has been permitted since the second century, the requirement cannot be absolute. The species are in this sense already quite mixed and can be used by all. Rabbi Avram Reisner writes: "Use of the product of hybridization is affirmatively permitted. In fact, many hybrids are presently on the market . . . those of separate species, which would be considered kilayim, the product agricultural and animal husbandry techniques honed before the advent of genetic engineering. No such product is banned. Indeed, this is not even a modern leniency, having its earliest source on the Tosefta."[23] The texts about the impermissibility of sowing two seeds together raise other questions.

By Misneaic times, it is already being queried on two grounds. First because of the technical impossibility of actually separating seed (not even possible in modernity, notes Wahrman)[24] and so clearly a problem in earlier eras that the number of unavoidable mixtures is specified (less than one rova in one seah or .5 liters to 13 liters, or 24 parts of the majority). Since genetic manipulation involves the transfer of DNA fragments invisible to the naked eye, it barely registers as significant in the classic halachic sense. For many food concerns, in fact, invisible entities do not even qualify as halachic issues—which is why, since the mideighteenth century, when the microscope revealed animal organisms living in every single gulp of water and bite of tofu, rendering everything nonkosher, it was ruled by Rabbi Yechiel Epstein that microscopic organisms were not to be considered a part of what constitutes eating (although, of course, they are intrinsic, as in yogurt, wine, vinegar). The idea of *batul b'shishim*, a concept that addresses accidental mixing of food as in cooking, states that less than one in sixty in a mixture is acceptable. But even *batul b'shishim* itself can be overridden in special cases—nearly all related to the use of the food for health or to avoid starvation or malnutrition.

There is some dissent, raising the issue of intention into the narrative. Some argue that if the intention is to change the mix, our acceptance of its smallness is not relevant, since the intention is to change the essential character of the food. But in the three decades since food processing was allowable by the rabbinic authorities, it has been established that such alteration is accepted. One can see this in a widening range of chemically altered versions of kosher cheese or "kosher bacon," made of soybeans, for example.

Textual Recourses in the Use of Genetically Altered Animals

For many, the genetic alteration of animals causes particular alarm, yet the process of genetic recombination in animal models is a long-held practice not only, as noted above, in animal husbandry, but also in basic medical research. For example, mice altered to carry genes known to cause human disease offer a close account of how the disease progresses and what drugs could be used to treat it. But what of the move to create animals for human consumption? What makes this particularly important is the elaborateness of rules that govern the raising, use, killing, and eating of animals in particular. That only certain sorts of animals are permitted for eating, and that the determinants of permissibility are phylogenetic characteristics, such as the shape of hooves, the existence of two stomachs for rumination, and so on, makes genetic alteration of physical attributes important.

Animal sacrifice was once the core of worship itself at the Temple in Jerusalem until its final destruction in 70CE. Even in modernity, Orthodox and traditional Conservative prayer books carefully recount the number, kind, and method of killing and presenting animals to the altar to be burned or to be eaten by the priestly class that attended the Temple. The liturgy of Yom Kippur serves as a narrative reenactment of the sacrifice and consumption of various animals, most famously the "scapegoat"—one who is sacrificed for atonement and the other who is driven into the wilderness. The biblical texts remind us that the way to seal a sacred covenant was to sacrifice a firstling animal, and this is one of the central narratives of the Torah, the New Testament, and the Qur'an, when it is recalled that Abraham kills a lamb and not his beloved son Isaac. Finally, since the most obvious part of the system of kashrut is the need to separate all milk products, dishes, and utensils from properly designated kosher meat, the attention to the body of the animal is especially freighted. Since a single flaw or lesion in the animal's lungs, for example, marks the difference between kosher designations (the term being *glatt*), can a small alteration in genes render an animal flawed?

The consideration of the use of GM and transgenes in the creation of animals for food raised specific concerns, as outlined in detail by Miryam Wahrman, from the Orthodox tradition. In the lay community, the problem of error, witchcraft and demons, the slippery slope arguments about Nazi eugenics, and references to Catholic natural law theory—all are a large part of the contemporary discourse of opposition to GMOs. Yet, in the rabbinic commentary, in all cases, the strong overriding concerns of *pikuah nefesh* (saving lives) and *v'rappo virape* (heal, surely heal) have consistently been used. The arguments that permit other types of GM food are used in this case as well. It is argued that genetic alteration of fish or in milk production can increase access to food. Also, since many alterations are used not directly for the animal's consumption but are used to create products then used as medication, commentators understand the genetic alternations as fundamental to scientific progress and permit this activity.

Interestingly enough, while some worry that genetic alteration is akin to the case of the sexual union of two species to create a hybrid (as in the creation of a mule, which is forbidden), others argue that this is not an accurate precedent case, since no mating is actually occurring. Additionally, many rabbinic sources consider the rules of bitual, a rule that states that a mixture retains the property of the majority component if the addition is less than one-sixtieth of the total. Since a DNA fragment, while able to replicate proteins that alter function, is still from one-trillionth to one-billionth of a gram,[25] it falls well below the halachic quantitative limits as it is far less than the amount stated as critical, and it is also entirely invisible. In fact, several commentators do not regard the transfer of genes as significant even if a gene from a nonkosher species is transferred into a kosher one. Such an opinion is based on another consideration within the tradition, *ayno raoey l'akhila*: a forbidden food that is unfit to eat ("even for a dog") loses its status as forbidden and may be used in any way.[26] Consider this definitive opinion from R. Nisson Shulaman, in his written responses for the Chief Rabbi of Great Britain: "Judaism does not regard genes as food, so the introduction of a gene or genetic material from non-kosher animals would not render the recipient non-kosher."[27]

These grounds may seem unduly technical, yet the law has always been understood and interpreted in this way, for even in the Gemora rabbinic authorities discuss precisely how much hemp can be mixed with linen to make the mixing of linen with wool permissible. In fact, while the yoking of two specific animals (ox and ass) is not permitted, other kinds of mixed yoking—including, interestingly enough for our purposes, the yoking of a man and an animal—are clearly discussed, considered, and permitted, rendering the argument that any genetic mixing would be forbidden simply incorrect. The rabbinic project of textual interpretation, in fact, even considers the case of how to use the stomach of a young animal, which clearly must contain milk—indeed, even its mother's milk, and they permit this, after reasonable efforts have been made to remove it.[28]

Here, one sees the effort to allow communities, for which every single part of an animal is valuable and necessary as food, to be able to use as much as possible of the meat in question. Here again, we understand that economic justice factors in the consideration of the limits of kashrut. In this text, the rabbis consider that axiom that "for everything that is forbidden, God has given us a way to use it" as a justification for the rabbinic interpretive praxis that can allow even this clear exception.

For Miryam Wahrman and others, the use of animals enhanced by genetics is an optimistic sign. She points to the animal that is "pharmed" or specially created to produce a certain useful protein in its milk or eggs.[29] The only limiting factor for Wahrman is the halachic caution not to cause unnecessary suffering (*tzar baalei chayim*). She is not alone in this view. This command, according to Bliech, exists not only on behalf of the animal, since we can kill an animal for food, but also to shape the consciousness of one's duty toward animals. As Bliech notes, we have dominion but not cruelty as a justification for our stewardship of the world. Thus, the act of creating an altered creature is permitted. Additionally, the "taking" of DNA from an animal is not, as members of the the focus group fear, a painful intervention. DNA exists in every cell and is easily obtained. Further, animals that produce milk with pharmaceutical proteins are not subject to any sort of unusual pain; in fact, because of their particular use, they are treated far more carefully than the average milk cow.

The Texts of Healing in the GM Debate

Nearly all commentators turn to sources of healing to support GM technology. This idea—v'rappo y' rappe-heal, "you shall surely heal," is a core source text. While skepticism about the market economy shadows many accounts, if genetic manipulation is done, as in our case, to produce cheaper, better, more humanly useful food, Jewish law would support such a use. Notes Kenneth Waxman:

> Halachah's relationship to scientific medical or therapeutic interventions to combat disease and prolong life, might very well provide a paradigm.... [G]enetic manipulation ... (and) the practice of medicine can be construed as suborning the natural order and contradicting the will of God. After all, in the natural way of things, people get sick and die.... Nevertheless, despite the possible support of certain biblical and Talmudic texts for a quietist orientation that accepts disease, as the will of God, and eschews therapeutic intervention; the Halachah has explicitly endorsed activism in this instance and in the practice of medicine. The Talmud explicitly deduces the physician's license to heal from the doubling of the Hebrew stem for "heal" in Exodus 21:19 (see Babba Kamma 85a). Furthermore, the practice of medicine is not only permitted, a *devar reshut*, but on the consensus account constitutes the fulfillments of a positive divine command; it is *devar mitzvah*.[30]

Voices of Caution and Their Interlocutors

While a close reading of the halachic reasons for accepting GM food clearly reveals the rabbinic support, especially for cases such as the work being done in the Pruess lab, there are voices that worry about the general ecological issues. Waskow, noted earlier in this chapter, is not alone in his concerns or audience. Unlike the unanimity among Orthodox and traditional commentators, some Conservative and Reform scholars, usually the most permissive in matters of food, law, and eating, in particular, find reasons for concern. Robert Gordis argues for a world in which humans are not at the center of creation: "Man [sic] takes his place among the other living creatures or the natural resources to be found in a world not of his own making, nor intended for his exclusive use."[31]

Conservative Rabbi Shammai Englelmayer notes the problem of *kilayim* may still apply, while Rabbi Joshua Finkelstien worries that GMO technology may create "unforeseen problems" (although how this is distinct from other nonmolecular interventions, such as the use of fertilizer or pesticides, is not specified). He notes that every intervention will drive corresponding changes in other species, a problem noted since Darwin and surely not limited to genetic intervention. Sholmo Rappaport notes that our lack of full understanding of molecular mechanisms may result in the creation of "genetic monsters." When some raise the specter of witchcraft, long a concern in Jewish law, the traditional and Orthodox authorities respond firmly that it is not and respond to such critiques by their Reform, Conservative, or "new age" colleagues by recourse to the texts. For example, Rappaport's contention is soundly refuted by Avram Steinberg:

> According to the Rambam (Guide for the Perplexed, III: 37), anything for which the association of cause and effect is established by accepted scientific rules and is dictated by natural logic does not constitute witchcraft; likewise, anything proven by experience, even if not prescribed by logic, is permissible. Even according to the opinion of those rishonim who maintain that witchcraft uses actual efficacious forces, the technology in question does not seem to be included in the prohibition of witchcraft. The Torah's prohibition of witchcraft according to this approach refers to the use of evil spirits or powers with destructive intent. (See Ramban, Ex 7:11)[32]

This is the case, argues Steinberg, even if acts seem to contain an "aspect of sorcery" in some way, for they were forbidden only because of their harm. Here he cites Shabbat 67b and Sefer ha Hinukh: "The Meiri, in his bet HaBehirah commentary to Sanhedrin 67b writes that "anything done as a natural activity is not included in the prohibition of witch craft. *Even if one knew how to create creatures without natural procreation, as is known in the books of nature, he may engage in this activity, since anything natural does not fall under the rubric of witchcraft.*"[33]

Indeed, the creation of "creatures" with de novo capacities via genetic engineering can even be salutary, as was noted above in our discussion.

Still, for some, the general public fears intrude into the halachic structure. Many others note the problems of unforeseen changes, or allergenic potentiation, or errors inherent in all scientific advances. As early as 1956, when such technology was in its infancy, Rabbi Emmanuel Jacobowitz cautioned against too rapid advances for safety reasons, calling for a moratorium on genetic engineering, a policy that was indeed followed voluntarily by the practitioners in the field until safety could be proven. "It is indefensible to initiate controlled experiments with incalculable effects on the balance of nature and the preservation of man's incomparable spirituality without the most careful evaluation of the likely consequences before hand."[34]

Indeed, Jacobowitz's caution has been not only carefully followed in Europe, but the development of such a "precautionary principle" has been written into codes of law. Yet, once established, the rDNA pharmaceutical interventions have been robustly embraced, with Israeli researchers at the forefront of many of these advances.[35]

Still, unlike the rapid acceptance of computers, cell phones, and wireless internet, whose increasing use brought increasing acceptance, the use of any technology in the making of food has become an ever more hotly contested problem. In 2004, the public was able to raise new and real concerns, and not easily dismissible ones, about even *conventional* food production, as stories of mad cow disease and contaminated wild fish filled the papers. Could we trust that any science used in agriculture was truly well intended? Or would marketplace pressures allow risky choices to be made? Finally, as many have noted in the consideration of other genetic research projects, our idea of what is "good" is shaped by culture, and our culture has turned rather alarmingly to valorizing largeness, sameness, and inoffensive taste in food. One is understandably drawn to ask: is there a place for the preservation of awkward heirloom tomatoes or fragile, unshippable peaches in a hungry and desperate world? Are our concerns about the aesthetics or grace of food rendered absurd by the bleak realities of famine in Nigeria or by the fact that even as I write this chapter, and as you read it, thousands of children will sicken and die of hunger? Was genetically modified corn the issue, or was any food, even organic lettuce produced in and shipped from California, even more problematic?

Safety and Policy Studies

Such was the state of the debate in July 2005, when a long-term, careful study was released by the World Health Organization. Here, the WHO explored the use of genetic manipulation and its impact on both humans and the environment. The findings of the report were startling. For despite fears of harm, "no health injury has been found, and there were many improvements in overall nutritional status. Additionally, in many fields that were monitored, the GM fields were more ecologically balanced, for when farmers use less pesticide, insects and wild plants were more evident."[36]

The United Kingdom also sponsored a multiyear study on GM agriculture that leads to similar conclusions, leading both the UK and WHO to conclude that no adverse risks can be said to exist for GM food production.

Texts of Justice in the GM Debate

It is my contention that in many debates about faith and science, a careful reading of texts, with broad inclusion of different interpretive possibilities found within the tradition, can be useful in finding good choices. It is in light of this contention that I have returned to the sources to locate what many will find counterintuitive and that in fact were a surprise to me as well as to the focus groups, my collaborators, and reviewers of this larger project.

But given the depth of the opposition to GMOs, especially among the progressive movement and liberal Judaism, I will make another claim: that there are more and different reasons, with more and stronger texts, to encourage the careful and thoughtful development of genetic research and technology. Indeed, it is these texts that allow me to return to the case with which I began and to respond to Dr. Pruess's question.

To conclude, let us now go to a different sort of reading, one in which we return to the original text of hunger, production, and limits. Let us return again to the text about mixing that so troubled the member of the Canadian focus group, and read it in its context.

Deuteronomy, Chapter 22

1. You shall not watch your brother's ox or his sheep go astray, and hide yourself from them; you shall in any case bring them again to your brother.

2. And if your brother is not near you, or if you know him not, then you shall bring it to your own house, and it shall be with you until your brother seeks after it, and you shall restore it back to him.

3. In like manner shall you do with his ass; and so shall you do with his garment; and with every lost thing of your brother's, which he has lost, and you have found, shall you do like-wise; you may not hide yourself.

4. You shall not watch your brother's ass or his ox fall down by the way, and hide yourself from them; you shall surely help him to lift them up again.

5. The woman shall not wear that which belongs to a man, neither shall a man put on a woman's garment; for all who do so are an abomination to the Lord your God.

6. If a bird's nest chances to be before you in the way in any tree, or on the ground, whether they are young ones, or eggs, and the mother sitting upon the young, or upon the eggs, you shall not take the mother with the young;

7. But you shall let the mother go, and take the young to you; that it may be well with you, and that you may prolong your days.

8. When you build a new house, then you shall make a parapet for your roof, that you should not bring any blood upon your house, if any man falls from there.

9. You shall not sow your vineyard with different seeds; lest the fruit of your seed which you have sown, and the fruit of your vineyard, be defiled.

10. You shall not plow with an ox and an ass together.

11. You shall not wear a garment of different sorts, like wool and linen together.

12. You shall make fringes upon the four quarters of your cloak, with which you cover yourself.

What is occurring here? The people are assembled, remembering their receiving of the law at Sinai. Moses rereads the law, making some changes from the reading in Leviticus. The verse that we are looking at is in the midst of a list of prohibitions, beginning with the case of restoration of lost property that is one of the proof texts for healing and medicine, for if you must return a lost ox or coat to your kinsman, surely, you would restore his lost health to him if you could! Then there is a series of confusions, including one that reminds us not to confuse genders and one that is the template for restraints on allowable hunting: you must not kill the mother bird along with her young or eggs, for that would end the process of generativity that allows the world to continue. Can there be a connection? To what sort of caution are we being advised? The final prohibition is logical: do not build a house unless the roof has a railing. After this is our verse. The prohibition on seeds is more specific: you have a vineyard, a permanent sort of field, and you cannot sow it with seeds. For the seeds cannot mix, fruit and fruit. Or is it that you can have a vineyard, and sow it with seeds of one kind, then another? It is not entirely clear, nor is the defilement explained. Then we are told of two other prohibited mixings: you cannot plow with an ox and an ass; you cannot wear wool and linen; you cannot look entirely like other peoples, but must be visibly marked, each corner of your outer garment, on the edges of your outer garment, marking your own edges. In later texts of the Tenach this biblical passage is recalled by the prophet Isaiah:

Isaiah

And when you reap the harvest of your land, you shall not reap to the very corners of your field, nor shall you gather the gleanings of your harvest.

10. And you shall not glean your vineyard, nor shall you gather every grape of your vineyard; you shall leave them for the poor and stranger; I am the Lord your God.

11. You shall not steal, nor deal falsely, nor lie one to another.

12. And you shall not swear by my name falsely, nor shall you profane the name of your God; I am the Lord.

13. You shall not defraud your neighbor, nor rob him; the wages of he who is hired shall not remain with you all night until the morning.

14. You shall not curse the deaf, nor put a stumbling block before the blind, but shall fear your God; I am the Lord.

24. Does the plowman plow all day to sow? Does he open and harrow his ground?

25. When he has prepared a smooth surface, does he not scatter the black seeds, and scatter the cumin, and throw in wheat

by rows, and the barley in the marked spot, and spelt along
its border?

26. For he is instructed properly; his God instructs him.

This is a curious rereading of the rereading. Here, the prophet considers the prohibitions of harvest and deepens them, linking their observance
not only to the sanctification of God's name but also to the need for social
and economic justice. The laws of harvest stress the point that the poor are
an important part of the productive process—the land belongs to God, and
the landholder is merely using a part of it for his family. He must set up
production so that the poor also have access to their part of the land—the
corners of the field—and the gleanings are their entitlement, not dependent
on the charity of the landowner. Can it be that the mixing of the seeds rule
is directed to the same end? Clearly, something interesting is hinted at by
the careful delineation of the fact that indeed, one can mix crops in some
way: here is the barley and there the spelt and here the wheat in rows, the
black seeds of sesame and the cumin thrown about? Clearly the instructions
for this order are also nonstochastic; they are given by God, and the order
of seed mixing is clearly a large part of the production valorized here as a
farm in line with the divine order of creation itself. The final rereading is
later, a midrashic one:

Midrash Rabbah—Deuteronomy VI:4

[I]f you have a field and you have gone to plough therein, the
precepts accompany you, as it is said, Thou shalt not plough
with an ox and ass together (Deut. XXII, 10); if you are about
to sow it, the precepts accompany you, as it is said, Thou shalt
not sow thy vineyard with two kinds of seed (ib. 9); and if you
reap it, the precepts accompany you, as it is said, When thou
reapest thy harvest in thy field, and hast forgot a sheaf in the
field (ib. XXIV, 19). God said: "Even if you are not engaged
on any particular work but are merely journeying on the road,
the precepts accompany you." Whence this? For it is said, IF
A BIRD'S NEST CHANCE TO BE BEFORE THEE IN THE
WAY, etc.

In this Midrash, we are back in the vineyard, and you have to carry your
Law with you.

Even in the most mundane and quotidian of tasks—farming—your
actions have to correspond to the law.

Midrash Rabbah—Numbers X:1

R. Levi said: All Israel's actions are distinct from the corresponding actions of the nations of the world; this applies to
their ploughing, their sowing, their reaping, their sheaves, their
threshing, their granaries, their wine-vats, their shaving, and
their counting. "Their ploughing?" Thou shalt not plow with
an ox and an ass together (Deut. XXII, 10). "Their sowing?"
(Num. 334) Thou shalt not sow thy field with two kinds of
seed (Lev. XIX, 19). "Their reaping?" Thou shalt not wholly
reap the corner of thy field (ib. 9). "Their sheaves?" And hast
forgot a sheaf in the field . . . it shall be for the stranger, for
the fatherless, etc. (Deut. XXIV, 19). "Their threshing?" Thou
shalt not muzzle the ox when he treadeth out the corn (ib.
XXV, 4). "Their granaries and their wine-vats?" Thou shalt not
delay to offer of the fullness of thy harvest, and of the outflow
of thy presses (Ex. XXII, 28). "Their shaving?" Neither shall
they shave off the corners of their beard (Lev. XXI, 5). "Their
counting?" Israel fixes their calendar by the moon, the nations
of the world by the sun. Moreover, when thou takest the sum
of the children of Israel, etc. (Ex. XXX, 12), and Take ye the
sum, etc. (Num. 1, 2).1 "To be mine": behold, ye are Mine. So
we have explained, "This is my beloved." And this is my friend
(S.S. V, 16). R. Judan, in the name of R. Hama b. R. Hanina,
and R. Berekiah, in the name of R. Abbahu, expounded: He
makes many friends for me.2 How? It is written, And have set
you apart from the peoples (Lev. XX, 26), i.e. like one who
selects the good from the bad; such a man selects and again
selects3; for any one who selects the bad from the good does
not go on selecting.4 So the Holy One, blessed be He, waits for
the nations of the world in the hope that they will repent and
be brought beneath His wings.

In this account, we see the fullest examination of the problem and
the hints at the elements of concern. The people's distinctiveness lies in
the covenant and the land—but the land is complex, for cultivation, unlike
nomadic wandering, requires estates, crops that take years to grow, gains and
losses, and the need for debt and forgiveness. Gleaning is a part of this, yet
to glean requires a particular sort of crop, and it requires a cessation and a
pause in the productive process. One cannot, as it is tempting to do in a
mild Mediterranean climate, simply keep sowing seed to overlap harvest on
harvest, for one must make room for the poor to glean in the interstitial,

liminal space "after the harvest" of one crop and before the fruition of the next. Thus the crops may be mixed if they do not compete and if the poor can glean without destroying the growth of the next crop. If the land "owner" felt that a poor person destroyed his crop, he might be tempted to shut her out—and he is entitled to first pick. But this production is limited and shaped, not only by his own hunger, but by the hunger of the Other, silently coming to interrupt the use of the land with her need and her right. One's self, one's daily work, a thing as small as a seed! All will matter in how the harvest—the core of the world itself—is made.

Belovedness, the covenant, and promises that make Israel and God lovers are interwoven quite literally in the rhythm of the text, with rules, borders, and limits: specificity. Later texts, Sifra 230 and Sifre, which comment on this issue, are even more explicit. The rabbis reread the spareness of the law into the richness of the commands toward the poor, linking the two so that ontological and moral meaning is made. Not mistaking, not carelessly throwing the very promise of the future—one's seed crop—around, but planting it with utter intentionality, is what makes the world acts word acts and hence tied to the sacred nature of human things. It is precisely the nonnatural, deliberative quality of this law that concerns our commentators.[37]

Conclusion: The Poor at Surrounding the Field

Such are the concerns, I would argue, that might guide a thoroughly defensible Jewish response to the problem raised by the specific case that Dr. Pruess brings and may serve as a general guide, in addition to the well-established responses from the scholars who define Jewish legal codes about GM food in general. It is not an answer to a science fiction (to pigs who chew their cuds or designer pets); it is an answer to a world right now, with the poor, right now, waiting their turn in our harvest. The textual midrashic answer is consonant with the halachic one: genetic engineering is not prohibited in Jewish law, and in many cases, is defensibly required if done with correct attention to the actual command one might think prohibits it—*kilayim*. For the cassava project is just such an idea: what do we owe the poor that they might harvest decently?

GM foods have extraordinary promise. In this way, they are like seeds themselves, for the entire thing could fail utterly, defeated unless watched carefully, nurtured through a risky start, and harvested in careful accordance to the laws of justice and responsibility. As we watch these beginnings, it is prudent to recall the midrashic admonition: *when we plow the field, you must take the law with you.*

Notes

1. See the Howard Hughes Medical Research Institute website, www.hhmi. org, for a full description of Professor Pruess's research projects, from which this quote is drawn.

2. Biocassava project on Gates Foundation website, 2005.

3. Other questions included whether it matters how or where food is produced, purchased, prepared, and eaten, and whether, according to Judaism, there are any foods that should be avoided because they are considered unacceptable.

4. Argues Miryam Wahrman, GM foods are, in the first world, a staple of our abundance: "About 60–70% of the food in supermarkets may contain genetically engineered ingredients, according to the Grocery Manufacturers of America. These foods include cereals, soft drinks, ice cream, chocolate and a host of other products. The reason is that about half the soybeans and a quarter of the corn grown in the United States contain genes from bacteria that make the plants resistant to herbicides or insects."

5. Mark Washofsky, "The Power to Cure," *Chicago Jewish News*, June, 2006, p. 15. I use this second printing deliberately to demonstrate that this opinion is not esoteric but is considered both authoritative and well known as well as well publicized.

6. As noted by Miryam Wahrman in "Curioser and Curioser: Tesuvah on Genetic Engineering," *Conservative Judaism* 52, no. 3 (2000), p. 197.

7. Arthur Waskow, *The Kosher Pathway: Food as God-Connection in the Life of the Jewish People*, http:// www.theshalomcenter.org, 2006.

8. Much of our DNA is highly conserved (which is why we are so similar to mice and yeast and why much of modern medical research is done on a few model animals).

9. Ruling by R. Chaim Pinchas Scheinberg, shlita, or Somayach international.

10. Miryam Wahrman, *Brave New Judaism: When Science and Scripture Collide* (New York: Brandies University Press, 2002), p. 211. My thanks to Dr. Wahrman for her exhaustive summary of the arguments from which I will cite the major ones relevant to our case.

11. Indeed, it is not uncommon for a kosher brand to lose this status if a violation is found.

12. Saul Berman, "Mitzot and Ethics." Speech given at the Edah Conference on Modern Orthodoxy, May 2005, Skokie, Illinois.

13. Waskow, *The Kosher Pathway*.

14. Berman, "Mitzot and Ethics."

15. Avraham Steinberg, "Human Cloning—Scientific, Moral and Jewish Perspectives," in *The Torah U'Maddah Journal* 9 (2000), Special Issue on Genetic Modification, p. 199.

16. Wahrman, *Brave New Judaism*, p. 213.

17. Any commandment except three—murder, idolatry, and adultery—can be abrogated to save or heal a human life.

18. Waskow, *The Kosher Pathway*.

19. Other things can be kosher, even acts thus declared, but the largest category surrounds food.

20. The subject of the modification of animals engages a different set of variables than the ones directly indicted in the case of biocassava.

21. Cited in Wahrman, *Brave New Judaism*, p. 196.

22. Ibid., p. 197.

23. Ibid.

24. Ibid., 213.

25. Ibid.

26. Ibid, p. 205.

27. Ibid.

28. For example, by flinging the stomach against a wall. R. Natan Levy, in a conversation in 2005, led me to this consideration in Chillin. It was taught in agreement with the first version of Rab's view: if the udder was cooked with its milk it is permitted; if the stomach [of a sucking calf] was cooked with its milk it is forbidden. And wherein lies the distinction between the two? In the one the milk is collected inside, in the other it is not collected inside.4

How should one cut it5 open?—Rab Judah replied. One must cut it lengthwise and breadthwise and press it against the wall. R. Eleazar once said to his attendant, "Cut it up for me6 and I will eat it." What does he teach us? Is it not [a clear statement in] our Mishnah?—He teaches us that it is not necessary to cut it both lengthwise and breadthwise.7 Or [he teaches us that this would be sufficient even for cooking] in a pot.8

Yaltha 9 once said to R. Nahman: "Observe, for everything that the Divine Law has forbidden us it has permitted us an equivalent: it has forbidden us blood but it has permitted us liver; it has forbidden us intercourse during menstruation but it has permitted us the blood of purification;10 it has forbidden us the fat of cattle but it has permitted us the fat of wild beasts; it has forbidden us swine's flesh but it has permitted us the brain of the shibbuta;11 it has forbidden us the girutha12 but it has permitted us the tongue of fish;13 it has forbidden us the married woman but it has permitted us the divorcee during the lifetime of her former husband; it has forbidden us the brother's wife but it has permitted us the levirate marriage;14 it has forbidden us the non-Jewess but it has permitted us the beautiful woman15 [taken in war]. I wish to eat flesh in milk, [where is its equivalent?]." Thereupon R. Nahman said to the butchers, "Give her roasted Udders."16

29. Wahrman, *Brave New Judaism*, pp. 190–92.

30. Kenneth Waxman, "Creativity and Catharsis: A Theological Framework for Evaluating Cloning," *Torah U'Maddah Journal* 9 (2000), p. 190.

31. Nachmonides, BNJ, p. 226.

32. Stienberg, "Human Cloning—Scientific, Moral and Jewish Perspectives," p. 202.

33. Ibid, p. 203. In both this quote and the one above, Stienberg is discussing not only genetic engineering, but also its farthest reaches—human cloning.

34. Emmanuel Jacobowitz, cited in Wahrman, *Brave New Jerusalem*, p. 225.

35. Technion Institute, Yearly Report, Hadassah Hospital, 2006. See also Hebrew University and Bar Ilan annual reports.

36. WHO Report on Genetic Engineered Crops, July 2005, Geneva.

37. For additional texts relevant to the sowing of mixed kinds of seeds and the breeding of animals of separate species, see Berachot 22A, Shabbat 139A, Shabbat 84B, Talmud - Mas. Shabbath 85A, and Genesis 30.

5

Some Christian Reflections on GM Food

Donald Bruce

Introduction

Various forms of Christian faith and practice are found around the world in many different cultures. The term *Christian* means many different things according to its context—the basic system of belief, the practice of a particular local or national church, or the general culture, which may also be called "Christian' " if Christianity is the major religion, although the churches may consider this latter a poor representation of their faith.

Within their core beliefs, Christian churches display an immense variety of cultural expression. This in turn leads to many different views and practices about food and about its genetic modification. A single definitive Christian account of GM food shared by the diverse billions of Christians of the world would be impossible. As a scientist and Protestant theologian working for a Presbyterian church in a rich industrialized country, my context for framing the issues differs greatly, say, from a Catholic subsistence farmer in sub-Saharan Africa.

With these important caveats, this chapter considers first how food relates to Christian belief and practice. The issue of the acceptability of GM food is then explored in relation to various aspects of our faith and from differing viewpoints, illustrated by examples and from the two focus groups conducted with Mennonite Christians and Seventh-day Adventists in Vancouver in 2005. The level of reflection varies greatly. Some European Protestant churches were proactive in investigating GM crops and animals already in the mid-1990s. The German Evangelische Kirche Deutschland (EKD),[1] the Dutch Multidisciplinary Centre for Church and Society (MCKS)[2] and the Church of Scotland's Society, Religion, and Technology Project[3] also explored the theological, ethical, and societal issues ahead of the public debate. In Canadian churches, there has been considerable recent study.[4]

Generally, however, churches have been slow to take up the issue, responding reactively to a debate happening external to them. With some notable exceptions, such as the Catholic theologian and former botanist Celia Deane-Drummond,[5] the Catholic and Orthodox traditions have given markedly less attention to GM food than, say, to human genetics issues. There has been no formal declaration from the pope on GM food as such. While the Orthodox tradition has considered ecology at many levels,[6] there have been few statements on GM food.[7]

Attitudes among Christians toward GM food vary widely from enthusiasm to outright opposition but often lie somewhere in between. Information about genetic modification of food comes to most believers mediated through the filters of secular media, government, the public relations material of the bioindustry and NGOs, local hearsay, and so on. Much depends on how one first heard about GM, and who presented the issues and with what bias. In countries where the debate is politicized and polarized into campaign claims and counterclaims, getting to the truth presents a serious problem for many Christians. They wish for trustworthy sources of information to weigh up the issues for themselves. Where churches have established a presence in local and national debates over GM food, they have often found themselves playing the role of honest broker in a debate where trust is in short supply.

Food in the Christian Tradition

Christianity does not have specific food requirements as basic tenets of the faith. This is because of the central Christian belief that salvation is a gift by the grace of God, which no amount of human effort, moral living, or religious practices can achieve. In Jesus Christ, God has come in person, living a perfect life, dying for all human sins of past and future, and rising from the dead to reconcile humanity to God. Through Christ, the believer enjoys a relationship with God and seeks to live a life consistent with God's ways, as revealed in the Bible and the Christian traditions handed down by the church. Because God has come to us, no further sacrifices are necessary.

The New Testament writings declare explicitly that there are no prohibitions on any type of food. Christ fulfilled all the dietary and ritual requirements specified in the parts of the Hebrew scriptures that Christians call the "Old Testament." Jesus taught that what defiles us is not any food that we eat but the sin already in our hearts, and in the gospel concludes, "In saying this, Jesus declared all foods clean" (Mark 7:14–23). Paul argues that food does not bring us near to God (1 Corinthians 8:8) and that "[t]he kingdom of God is not about eating and drinking but righteousness, peace and joy in the Holy Spirit" (Romans 14:17).

Excess of food and drink are inconsistent with holy living. Christians are to be filled with the Spirit of God instead of being out of control in drunkenness or addiction. Some Christians choose to be vegetarians, out of concerns about animal welfare and environmental resources or believing that God's original intention for humans was to eat plants (Genesis 1 and 9), eating animals being a later concession. A few groups within the broad umbrella of Christianity advocate the adoption of some Old Testament dietary stipulations as God's best plan for humans. For example, Seventh-day Adventists are forbidden from eating pork and certain other meats and fish and dairy products, but such requirements are unusual and are not accepted by the majority of church traditions.

Catholic and Orthodox traditions practice fasting at certain times, when only food taken from the earth is eaten, not meat. Fasting is an aid to spiritual awareness, to help focus away from daily affairs and habits, but is not done to escape from the body or from desires or appetite as such. The bodily incarnation and resurrection of Jesus Christ fundamentally emphasize the value God puts on human physicality. As an Orthodox scholar expresses it, "[b]y recovering a proper attitude to our bodies . . . we also recover a right attitude to the creation as a whole. We are helped to value each thing for its self—not just for the way in which it serves our own ends. Fasting, so far from being world-denying, is in reality intensely world-affirming. . . . It is God's world, a world full of beauty and wonder, marked everywhere with the signature of the Creator, and this we can discover through a true observance of fasting."[8]

Sharing a meal is important in Christian family life and in informal Christian fellowship and teaching. In the Mennonite focus group, this was expressed as follows: "We use food to show love. . . . Just as Jesus used the very simple act of eating with people to express love and compassion, I think, we use food to express a lot of things that we maybe don't always have the words for." Jesus' example of eating with outcasts is significant for many Christian groups as a call to show hospitality to outsiders. Several members of the Mennonite group stressed the creativity of food that is handed down in recipes, rather than ready made food. One said, "I use the term *slow food*. . . . It's the preparation of the food. . . . Food is more than just utilitarian—it's a spiritual undertaking of preparing food and eating. Growing up in my Mennonite Christian tradition, you couldn't eat food without praying. . . . You sit down as a family unit to eat. You don't eat and grab. It's a time to come together and focus and to offer thanks for that gift."

Finally, food is used representatively in the central ritual common to Christians everywhere, known as the Lord's Supper, Holy Communion, the Eucharist, or the Mass. A mouthful of bread and wine is consumed in a service of worship that focuses on two symbolic acts that Christ performed in the course of the last meal with his disciples before his crucifixion. The broken bread represents his body, broken on the cross for humanity, and the shared cup of wine represents his blood, shed to forgive the sins of the world, the sign of a new covenant between God and humankind. The significance is not the food itself but its symbolic use as a representation, reenactment, or memorial of the death of Jesus Christ, as an act that unites all believers and as a foretaste of the eternal banquet that is a metaphor of heaven. In many traditions, much indeed is made of the fact that the bread and wine are ordinary fruits of God's creation, adapted by human skill and offered back to God as tokens.

Genetically Modified Food as an Intrinsic Issue

General

The first question to be asked in relation to genetically modified food is whether it is intrinsically right or wrong to create crops or animals with genes from other species with which they could not normally breed. Many churches that have examined GM food in depth could not find a substantial theological case to say it is *intrinsically* wrong, although some groups have advanced arguments that would prohibit it absolutely or conditionally. More often concerns and objections raised by the Christian faith are about the consequences of modification or to its social context, in relation to important theological principles.

Biblical Data

Christians look to the Bible for foundational principles and particular commands. Not surprisingly, few texts relate directly to what we now call "crop" or "animal genetics." Selective breeding is practiced by Jacob, from whom the breed of Jacob sheep gets its name (Genesis 30). Members of the Seventh-day Adventist focus group observed that since God created the different creatures "after their kinds" (Genesis 1), we should not move genes to unrelated species. The *Engineering Genesis* study,[9] however, was inconclusive as to whether the Hebrew distinctions of "kinds" necessarily equates with current biological notions of species or whether such distinctions were meant to prohibit admixtures. The important case of Solomon riding King David's mule as a sign of his succession to the kingship over God's people (1 Kings 1:32–35) suggests that the mixing of kinds is not absolutely proscribed.

Two passages contain ordinances that prohibit mismating cattle, sowing two kinds of seed in a field, or making a garment out of two kinds of material (Lev. 19:19 and Deut. 22:9–11). These are not normally seen as a biblical sanction that the genetic modification of crops or animals is proscribed by God. As observed above, practical stipulations about food and agriculture in this section of the Hebrew scriptures are not binding to Christians. They would not form the basis of a major doctrine unless comparable texts elsewhere indicate an important moral principle behind the specific command. Thus the precept "Do not muzzle an ox while it is treading out the grain" (Deut. 25:4) correlates with wider biblical principles that legitimate uses of animals for food or economic gain must be moderated by care for their welfare.

It is hard to see here what would be the principle in verses about mixed sowing, which seem to have no parallels and no clear context to interpret them. A study by the UK Evangelical Alliance[10] commented that "the prohibitions have a primarily cultural referent. Their purpose is principally to maintain Israel's distinctiveness vis à vis the rest of the nations. It does not exclude the possibility that these laws also reflect the need to preserve natural distinctions; merely that this motivation is not explicit within the text. . . . We have thus found little substantial biblical evidence with which to support an intrinsic objection to genetic engineering as such."

GM Seen as a Reasonable Human Intervention in Creation

This leads us to consider genetic modification in the light of broad theological principles. The first of these relates to how far God's creation is open for humans to modify and in what ways. Historically, Christian thinking has generally seen intervention in the natural world as ordained by God in the creation ordinances that grant humans dominion over all the rest of creation (Genesis 1:26–28). This has been variously interpreted at different times and contexts—overcoming the threats presented to vulnerable humanity by the natural world, recovering the order of creation lost in human rebellion against God, or harnessing the forces of nature to human ends.[11]

The latter has been much debated in recent decades in the recovery of a theology of creation that reasserts the importance of our obligations to God in caring for creation. This is implicit throughout scripture but was largely neglected under the influence of secular ideas of human mastery over nature as though we owned it, accountable to no one.[12] The creation is God's and only given in trust to humans. We are accountable to God for our use of it. Dominion is limited by God's laws, which require humans to restrain their activity out of respect for the rest of creation, love for our neighbor, and care for the disadvantaged. In the recovery of elements from Celtic traditions, creation is once again seen as a consummate work of art that points to God and is to be treasured with wonder as much as it is a tool to be used.[13]

How then is GM food to be viewed in the light of the human relationship to creation? A European church study acknowledges that part of the expression of God's image in humanity lies in humans' ability to shape creation to meet our needs for food and sustenance.[14] There seem no a priori grounds why GM should be selected for special objection, when compared with the myriad ways in which humans intervene in creation.

Since antiquity humans have cultivated crops, domesticated wild animals, and bred food animals and plants to their own ends. Today's cereals and cattle are very different indeed from their original wild forms. Humans have delved into the interstices of the matter of creation. In chemistry we have violently changed the chemical structure of rocks and other natural materials such as minerals to create new compounds and alloys, such as the bronze of Solomon's temple or titanium for hip replacements. In physics we have manipulated the fundamental forces of creation, for instance to channel electrons to generate electricity. We have invented means of propulsion, energy, calculation, communication, and vaccination that have changed our world in ways unimaginable to our ancient forebears. While the manner and extent of such interventions may raise concerns, Christians today do not normally argue that the interventions themselves are wrong before God.

When so many other fundamental changes to matter are regarded as commonplace, it would then seem arbitrary and indeed illogical to declare that molecular genetic intervention in crops and plants is intrinsically wrong. If God has given humans the skill to do so, and no clear biblical prohibition has been ordained, it is argued that we ought not to reject the basic idea of altering food crops and animals by moving a few genes across species.

Creation can also be seen as filled with possibilities, which human beings are to develop, fashioning into new forms the potential of what God first created. Creation is not a static entity. God did not call humans primarily to be rangers of a nature conservation reserve. Science and technology are proper human activities within this God-given mandate to make advances in the human condition. In *Engineering Genesis* the authors observed, "In as much as it may preserve crops from pests or drought, help prevent famine, and provide humans with healing therapies for disorders and disease, the genetic modification of non-human life could therefore be said to be an intrinsically appropriate use of science and technology for the preservation of human life from the threats which the natural world frequently presents to it. . . . Christ's command to love one's neighbor as oneself is hereby expressed in helping and enriching the life of others."[15] But it also acknowledged the considerable proviso, as we shall see, that in the pursuit of technology we do not become proud of our own wisdom or fail to practice due foresight, justice, and proper respect for creation.

GM Seen as an Improper Human Intervention in Creation

Some Christians do, however, make inherent objections against GM food. The primary concern is that humans are in some sense "playing God" in wrongly changing what God has created. While playing God is not a strict theological concept, it expresses a perception of usurping the creative preroga-tive of God by doing something that belongs to God alone, taking on a role that is not ours to have. It implies that humans are presumptuous or foolish to think they can improve on the wisdom by which God has created all living things and the interlinked ways they grow and flourish.[16] For example, one member of the Adventist focus group said, "God created this world. He cre-ated the food that is safe and good. Who are we as human beings to try and play God and create new things that we don't know anything about really?" Much was made in this group of the importance of what is seen as "natural." This was expressed as doing "something that could not happen in nature," like mixing plant with animal or hybridizing plants to make crosses that only humans could bring about. For some members of the Mennonite focus group, transspecies genetic modification "raised the red flags" as a human interven-tion that did things that could never occur in a natural system.

The use of naturalness in any absolute sense is notoriously problematic in ethics, as the *Engineering Genesis* study discussed.[17] What is considered natural is highly relative to time, location, belief, aesthetics, and culture. Arguments from nature are seen as theologically ambiguous. The popular assumption that nature knows best is viewed with caution, as a notion derived more from questionable naturalistic or pantheistic assumptions about nature than Christian theology.[18] We live in a morally fallen context; what we call "nature" or perceive today as natural is not necessarily a good guide to what God intended. Playing God can also have positive connotations.[19] Humans were ordained by God to rule on God's behalf over creation. In this capacity we have made innumerable far-reaching changes to God's creation, which would not have happened naturally. Thus it is not sufficient to assert that GM is self-evidently unnatural and other interventions are not, without further elucidation of what is meant.

What Is GM Changing?

Behind the belief that in some sense genetic modification violates God's creation seems to be an assumption that to change a living organism genetically is to change the organism in itself. To alter something living in this way is seen as different from chemical or physical change to inanimate matter, or cutting down a tree, or eating a pig. This poses a question. If we do not see an inherent problem in having dramatically changed by selective breeding most of the organisms we use for food from what was "natural," why should molecular genetic change be unacceptable? To what extent is GM a natural extension of breeding or a disjunction? If it is a disjunction, is it necessarily wrong for humans to do genetic crosses that could never have happened in the way God has allowed evolution to develop?[20] Genes are natural, and large numbers are shared across many species. A small degree of genetic change across species seems to occur in nature.

One member of the Mennonite focus group made a distinction between the natural occurrence of "Heinz 57" varieties of dog and the changes across species that genetic engineering implied. GM was seen as modification of nature, to do something quite unrelated to that species, contrasted with selective breeding where selections were made *within* nature, which remain related to the species. In the Adventist group, to introduce a foreign gene to a tomato changes it unacceptably from what God intended. What God created naturally is something humans should not intervene in.

This raises another key question to understanding GM theologically. What is the nature of a living organism, and in what way does genetic modification alter it in ways that are acceptable or unacceptable? One view is that, if genetics tells us that organisms are characterized by the detail of their genetic blueprint, to add a foreign gene would violate what God created. Daffodil genes do not belong in rice because that is not what rice is. If God created a sheep with a certain genome, adding a human gene to produce medically useful proteins in the sheep's milk is inappropriate.

The alternative view argues that this is too a reductionist assumption about the nature of organisms, derived more from science than scripture. A biblical understanding is that organisms are multidimensional and cannot be satisfactorily described by genetics or any other single aspect. What is important about GM is what happens to the organism as a whole. To modify a crop or animal to make it infertile might be unacceptable, depriving it of a God-ordained role fundamental to itself. But if it did not cause harm to the organism, it would not be an inherent violation of the rice plant to add daffodil genes to increase the amount of a vitamin A precursor in rice nor to add a human gene to a sheep to make pharmaceutical proteins.

The UK government Polkinghorne committee, chaired by a Christian minister, examined whether certain types of genes or interventions might have a "moral taint" in religious contexts. Would it be cannibalism to eat beef with a human gene, for example? It was argued that because genes are copied millions of times in the process of performing genetic modification, no one would consume a gene that had actually been part of another human.[21] The logic seemed too reductionist for many. The "humanness" of a gene lies more in where it came from and the information it conveys than whether the particular atoms of the genes in the mouthful on my fork happened previously to have been in a human being or a bacterium.[22]

Concerns about GM because of the changes it causes to the intrinsic nature of an organism arise more with animals than with plants.[23] This was extensively discussed in *Engineering Genesis*.[24] Most farm animal GM applications have hitherto been novel medical uses, such as pharmaceuticals in sheep's milk. Apart from fish, most genetic traits of interest to food producers can be altered more easily through selective breeding, although new methods of modification may change this in future.

An insight of the Mennonite group was that selective breeding is acceptable because one stays within the immediate gene pool of the species. This was partly a perception about relative risks, but partly also it attributed a moral significance to the relatedness of species within God's creation. In such a context, skeptical Christians might be more comfortable with crop GM where the transgene came from another plant, especially one not too far removed, than with crossing kingdoms such as putting a fish gene in a strawberry to make it frost resistant. The latter would seem more a violation of categories.

GM as Technology Gone Too Far … or Not Far Enough?

Some Christians associate genetic modification with the trend to an overly technological agriculture and food production. The strawberry example illustrates another perception that "unnatural" expresses a sense that technological intervention has gone too far in reordering nature to human ends, like the demand for perfectly shaped colored fruit and vegetables. Some focus group participants commented, "I avoid buying apples in Vancouver because they're nothing like what I had when I grew up," and, "I like to have foods that are naturally grown because I am scared of the chemical additives." On occasion, this may be a somewhat romanticized view of the past. In much advertising, "natural" tends to portray an idealized version of "what my grandfather knew."

Questions of hubris and right and wrong uses of technology under God emerge as a better Christian yardstick than what is natural. What does God intend humans to do with creation? In his 2000 homily for the Jubilee of the Agricultural World, the pope said that "[i]f the world of the most refined technology is not reconciled with the simple language of nature in a healthy balance, human life will face ever greater risks, of which we are already seeing the first disturbing signs."[25] In the Mennonite focus group a view was offered: "I believe that God created us to exist in harmony with each other and with our environment. And part of that is eating and consuming in ways that are more natural and less manufactured, because I think God created things for a reason, and if we got back to things that were more natural, I think that would be more healthy for us, and also more conducive to right relationships with each other."

The context in both cases is the notion that the moral fall of humans has also harmed nature. The redemption of creation by Christ implies that Christians should work to restore creation, as far as may be, to what God intended. There are two opposing views of this, however. For the Mennonite discussant, the implication was that conventional technology in the food chain had taken us in the wrong direction, and GM implicitly follows suit. Mennonite insights into creation stress humility and respect for what has been created, a recognition of the fallen state of things. A tension is perceived between God as creator and human beings when we start to play creator ourselves. Conversely, a line of reasoning especially articulated by Francis Bacon in the sixteenth century, with roots in the teachings of the reformer Calvin, sees technology in the opposite sense. Human skill in the sciences is God's providence to be used in nature to overcome the effects of the fall.[26] For many Christians it thus becomes a matter of judgment as to whether GM is more hubris or a right use of talents.

One intriguing response comes from Amish farmers, a closed Mennonite Christian community in the eastern United States. Although perceived as antimodern, their assessment of technology is complex, posing questions from unfamiliar angles. As well as biblical concepts one of their key criteria is the effect a technological artifact or system would have on the relationships integral to their practicing Christian community. This leads to some highly original interpretations.[27] For example, cars and tractors are rejected within the community, but you may hire a car to visit a distant sick relative or use a tractor engine for milk cooling. But they will adopt a technology if they are convinced of the benefit. They tend to farm organically but may also use pesticides. Some have grown trial crops of GM potatoes and a GM nicotine-free tobacco, with the aim of making the community more productive.[28] The implication from within such a sensitive community is that GM need not necessarily be done out of hubris or lack of care either for the creation or our fellow humans.

GM Food and Risk

A second group of issues for Christians is that of risk. As discussed above, part of the concerns raised by some about genetically altering life forms that God has created is that humans are not God. We have God-given capacities in invention and creativity, but we remain finite in our knowledge and morally fallen in our understanding. As a result, some Christians argue that humans do not have the wisdom and foresight to know the implications and outcomes sufficiently to make changes as far reaching and self-propagating as the genetic modification of other species. Other believers, however, consider that GM should not be regarded as any more risky than other interventions, such as selective breeding, provided appropriate testing and monitoring are done.

GM and Selective Breeding

For the Adventist group, genes or proteins introduced into a plant from another species were seen as having unknown effects on the body. Some of the Mennonite group thought that there would inevitably be negative consequences, even from beneficially intended GM applications such as reducing pesticide use. They cited unintended gene flow by the pollen of GM canola reaching non-GM varieties of canola in the vicinity. This was seen as a consequence of genetic change that could not happen in natural systems. This line of reasoning assumes that within relatively similar structures and systems there is less risk of something unusual and damaging happening as a result of introducing a "foreign" gene. Gene and protein interactions within species are highly complex. If one takes a gene identified with a discrete function in a fish and transfers it into a completely different class of organism such as a strawberry, it presumes on too many unknown factors that it will work in a comparable way, with no odd side effects. Two theological assumptions are frequently made or implied—that what is natural is better, and human artifice will tend to make things worse, and that the distinctions God has created in the species we have today embody an intrinsic element of safety.

The contrary argument is that all such modifications require rigorous testing over a period of time to identify and avoid just such issues. The same applies to conventional breeding, where there are substantial unknown factors at each new mating. There are enormous uncertainties in crossing within a species, because one is mixing not two or three genes but half a genome. For example, some highly toxic potatoes were unknowingly produced by normal selective breeding. Poultry breeding has been taken to extremes in that specifically enhancing growth rate caused significant animal welfare problems. Nine-week-old chickens are spectacularly larger than they would have been forty years ago.[29] To stay within the gene pool of the species is in itself no guarantee of safety or the welfare of the organism. It is not clear that genetic modification is always and by definition riskier than breeding within the species. The biological evidence implies that God seems to have allowed elements of risk in breeding within species as well as across them.

Exercising Precaution ... but How Much?

In the face of the uncertainties and controversy, recourse is often made to the precautionary principle.[30] Here again, such a variety of interpretations of this exist that its use is not decisive. It remains a matter of judgment whether a given circumstance calls for precaution and at what level this should be implemented. A United Church of Canada resolution observed that "a sense of proportionality must come into play—i.e., to the extent that potential harms to health or ecosystems may be irreversible, irremediable, or uncontrollable, there is a greater need to exercise precaution."[31]

One approach is exemplified by the Church of Scotland.[32] Having considered the available evidence, it concluded that because the risks of different applications of GM food varied greatly from application to application, a blanket moratorium on all applications was inappropriate. On health risks, it stated, "[a]t present, provided tests and regulations are adhered to properly, GM seems no more likely, in itself, to be a cause of major health problems than any other way of growing food." Nonetheless, "it would be irresponsible to relax the principle of precaution too far. It also seems appropriate at present to concentrate on applications that are restricted in scale, and which confer strong human or ecological benefits, and to set up long-term monitoring to see that there are no serious unintended effects over extended periods of time."

A very different view was taken in a brief Catholic reflection on GM food by two Jesuit theologians based in Zambia, first presented in Rome in 2003.[33] For them the precautionary principle is "more than a temporary scientific safeguard." It is a fundamental ethical norm, a "call to humility before the awesome goodness of God's creation" that should guide our evaluation of genetic engineering. They cite as justification a papal statement:

"We are not yet in a position to assess the biological disturbance that could result from indiscriminate genetic manipulation and from the unscrupulous development of new forms of plant and animal life." The authors consider that genetically modified organisms are indeed a case where biotechnology is indiscriminately and unscrupulously applied, by being inadequately tested. Drawing on the controversial writings of Mae Wan Ho, they see GM methods as utilizing "very powerful unnatural laboratory techniques" and so posing serious risks to humans. They cite various unexpected harms from other fields of technology (DDT, CFCs, and thalidomide) and conclude that we "should not adopt a technology that is still inadequately tested."

Arguments from precaution are temporary positions taken in the light of present uncertainty, but anticipate eventual resolution. The critical question is what, if anything, would make one certain, either one way or the other? Deane-Drummond considers a different approach based on prudence and wisdom derived from the medieval Catholic theologian Aquinas.[34] She puts a stress on both the explicit and hidden motives of those involved, the responsibility and liability toward other actors, and how far the modification is directed to the good to which the creature is ordered. But it still does not tell us a priori whether a particular modification is acceptable or too risky.

Wisdom is not only a result of cold hard reasoning. It involves a broader sense of discernment, but it should not be reduced to feelings only. Criteria are needed to decide when one has got sufficient information to make a decision one way or the other. Jesus tells the parable of the unfruitful fig tree, which the owner wanted cut down. But the husbandman said, "[G]ive me a year and I'll give it especial care; if it's still unfruitful, then cut it down" (Luke 13:6–9). Genetic modification lies outside the normal experience of most Christians, and in trying to make sense of it they usually have only partial information. Much therefore depends on what, and whom, they take as authoritative sources of scientific evidence and interpretation. In Catholic debates on the issue, the Zambian study chose mainly critical sources, but a Pontifical Academy of Science conference[35] was criticized for hearing mainly positive evidence.[36]

The most conservative interpretation of the precautionary principle would insist that no genetically modified crops should be used until it can be demonstrated that no harm could result. The Evangelical Alliance study argues, however, that such a response to risk is out of step with the way God has ordained the universe and our human calling.[37] "The Bible makes plain that since humanity fell, there has never been any such 'golden age,' and that what is 'natural' is no guarantee of safety." Risk is not therefore something rudely thrust upon us by scientists to disrupt our Arcadian existence but is inherent in the way God created the world in which we are placed.

Since God made humans creative but not omniscient, we cannot demand a level of certainty from technological endeavors that God has not given human beings to have. To do so might even be seen as a kind of idolatry.

GM in the Light of Power Structures and Economic Drivers

A concern for justice, for the poor and the disadvantaged of society, is of great importance to Christians and is arguably the strongest concern raised in Christian assessments of GM food. This is closely tied to the questions of indigenous versus scientific knowledge and to issues of justice and power in both the developing and industrialized worlds.

GM and the Developing World

The Zambian study especially examined GM crops in the light of Catholic social teaching on these issues.[38] It represents a fairly typical example of the issues that are articulated by small farmer groups opposed to GM crops, stressesing the negative social, environmental, health, and agricultural impacts of foreign technology on the majority of small farmers in Zambia. This line of thinking links GM food to problems of injustice, colonialism, and the monopoly of foreign transnational corporations. It was influential on the controversial decision of the Zambian government to refuse unmilled U.S. GM maize as food aid during a famine in 2002. In that sense, this was an example of GM food being seen as taboo, out of its cultural context.

The study also raised questions about the validity of knowledge. Indigenous holistic knowledge and local technology are set against knowledge brought in from outside as a technological solution. The one is seen as holistic based on popularly collated knowledge; the other is seen as abstract and formulaic, based on scientific reductionist knowledge. There is a disjunction between scientific and cultural representations of nature. There does not appear to be any specifically Christian reason to favor the one or the other, as knowledge. What drove the authors, however, was the bias toward the poor in Christian teaching, which leads to a deep concern about the ownership and control of the knowledge. The agents bringing the technology are foreign and not under local control and jurisdiction. They have their vested interests far away from the local situation. They are therefore not accountable to the local farmers but to distant shareholders or governments. The guardians of local knowledge are part of the local culture and accountable to it.

Christian theology is sanguine about any knowledge, in light of the fallenness of human beings. A corrective is certainly needed to the presumption that Western science automatically trumps all other forms of agricultural knowledge. But indigenous knowledge should not be idealized instead. Just as science can be disfigured by hubris, indigenous knowledge can also be misled by sin in its own situation. Traditional methods have changed the creation, as all human activity does, and in many cases have caused damage.[39]

Christian-based aid agencies are divided about whether GM crops have a positive role to play. There is great uncertainty as to whether many of the GM applications invented for North American bulk commodity farmers are relevant to the needs of African subsistence farmers and their equivalents round the world. Different campaign groups cite positive and negative anecdotal evidence, often depending on their own viewpoint. What is not well discussed is the role of indigenous research within the country and international research geared toward developing countries that is not funded by companies, for which Christian theology would have more sympathy. While most Christians would agree that the prevalent vitamin A deficiency is best addressed by a proper diet, the profound problems of poverty are not easily solved. Pragmatically, GM rice with enhanced vitamin A precursor may have a role to play. But few such applications exist, and fewer still may be adopted in poorer countries, because they do not have the finances that multinational companies can command to implement their globally marketed products.

Animal Cloning and Food

An example of justice-motivated concerns from the industrialized world is raised by animal cloning. The original reason that Roslin Institute scientists near Edinburgh began nuclear transfer cloning was to use cloning as a tool for more effective genetic modification to produce pharmaceutical proteins in sheep's milk. Standard methods of adding DNA were random, very inefficient, and required using many animals. Nuclear transfer cloning offered a way to grow animals from cells that had been successfully modified and tested. In the event, although a better success rate was achieved, the cloning technique led to serious animal welfare problems. The Church of Scotland followed these issues closely and expressed concern when a U.S. company proposed cloning a prime bull at the top of the breeding pyramid for use in commercial beef production:

> To clone animals routinely for production is treating animals too instrumentally, given that natural methods of breeding exist. A case might be made if cloning helped spread a genetic modification to combat an animal disease like foot and mouth, but mere commercial production or supermarket efficiency are not enough

to justify cloning. Copying the complete genetic blueprint for efficiency's sake carries a factory mass production mentality too far into animal husbandry. Our fellow creatures are more than identical widgets on an assembly line. . . . We may use them, but we also need to remind ourselves that they are God's creatures first. Given the abuses which a commercial drive has led to in some areas of animal production, cloning is a place to draw a line.[40]

Conclusion

Many studies conclude that to modify food crops or animals genetically across normal species barriers does not *of itself* break any explicit biblical command or primary theological doctrine, such that all Christians should on principle refrain from eating GM food. It is strongly debated among Christians, however, whether or not GM food *as we currently have it* violates other theological principles. These include the degree and nature of permissible human intervention in God's creation, care for creation and its creatures under God, wisdom with regard to risk-taking, justice toward people, and concern for the poor and about undue power in large companies and governments. A Christian view of God, humanity, and creation seeks to hold in tension the notion of intervention to change the natural world and the notion of respect for and conservation of its vital features, creatures, and ecosystems. In the light of this, there is undoubtedly unease at the potential risks of unfamiliar modifications of life forms. Allied with this are holistic and relational attitudes toward food, traditional and local agricultural practices, and the link between human communities and the land. Other factors that lie beyond the immediate scope of the acceptability of modified genes involve debates over physical and intellectual property,[41] public accountability, and global responsibilities.

For some Christian groups or individuals GM food does represent unacceptable genetic intervention in principle or for one or another of these reasons. GM food may attain a taboo status, especially where it is taught as such and accepted without much discussion. Many Christians, however, do not see GM as unacceptable on Christian grounds; they welcome its applications and future potential and grow and eat its crops. The strongest concerns relate less to unacceptability than to uncertainty and suspicion in the face of unfamiliar and powerful technologies, in the hands of organizations we do not trust and who are not accountable to us. Such concerns apply widely to issues that go beyond genetic modification, but the GM food debate has brought them into renewed prominence as few other issues have. It is a debate in which the churches have a key role to play.

Notes

1. EKD, *Einverständnis mit der Schöpfung* (Gerd Mohn: Gütersloh 1991, 1997).

2. MCKS, *Biotechnologie: God vergeten?* (Utrecht: Multi-disciplinary Centre for Church and Society, 1997).

3. A. Bruce and D. Bruce, eds., *Engineering Genesis* (Earthscan: London 1999).

4. United Church of Canada, *Making Wise Choices: A Report on Genetically Modified Foods* (Toronto: United Church of Canada, 2005).

5. C. Deane-Drummond, *Aquinas*, "Wisdom Ethics and the New Genetics," in C. Deane-Drummond, B. Szerszynski, and R. Grove-White, eds., *Re-Ordering Nature* (Edinburgh: Clark, 2003), pp. 293–311, and also references therein.

6. A. Belopopsky and D. Oikonomou, eds., *Orthodoxy and Ecology Resource Book*, Syndesmos, Orthdruk (Bialstok, Poland: Orthodox Printing House, 1996).

7. World Council of Churches, *Official Church Statements or Documents on Bioethics and Biotechnology*, April 2004. Accessed June 24, 2006 from http://www.wcc-coe.org/wcc/what/jpc/biodocs.html (Geneva: WCC, 2004).

8. D. Oikonomou, "Christian Fasting," in *Mozaik 2* (2002), pp. 16-18.

9. Bruce and Bruce, *Engineering Genesis*, p. 94.

10. Bruce and Horrocks, *Modifying Creation* (Carlisle, England: Paternoster, 2002), pp. 52–54, 168–75.

11. P. Harrison, "Subduing the Earth: Genesis 1, Early Modern Science and the Exploitation of Nature," in *Journal of Religion* 79 (1999), pp. 86–109.

12. See R. Page, *God and the Web of Creation* (London: SCM, 1996), and L. Osborn, *Guardians of Creation* (Leicester: IVP Apollos, 1993).

13. I. Bradley, *God Is Green* (London: Longman and Todd, 1990).

14. Conference of European Churches, *Genetically Modified Food* (Geneva: Church and Society Commission of the Conference of European Churches, 2002).

15. Bruce and Bruce, *Engineering Genesis*, p. 87.

16. Ibid., pp. 84–85.

17. Ibid., pp. 88–92.

18. Bruce and Horrocks, *Modifying Creation*, pp. 49–51.

19. E. Schroten, *The Philosophical Basis of Ethical Issues in Genetic Engineering*, Methodist Church Conference "Theological and Moral Concerns Involved in Genetic Manipulation" (Luton: Industrial College, June 1992), pp. 19–21.

20. M. Reiss and R. Straughan, *Improving Nature* (CUP: Cambridge, 1996).

21. Ministry of Agriculture, Fisheries, and Food, *Report of the Committee on the Ethics of Genetic Modification and Food Use* (Polkinghorne Committee) (London: HMSO, 1993).

22. M. Reiss, "Is It Right to Move Genes between Species: A Theological Perspective," in C. Deane-Drummond, B. Szerszynski, and R. Grove-White, eds., *Re-Ordering Nature* (Edinburgh: Clark, 2003), pp. 138–50.

23. D. Heaf and J. Wirtz, eds., *Genetic Engineering and the Intrinsic Value and Integrity of Animals and Plants*, Proceedings of the Ifgene Workshop, Edinburgh, 18–21 September 2002 (Dornach, Switzerland: Ifgene, 2003).

24. Bruce and Bruce, *Engineering Genesis*, pp.127–58.

25. John-Paul II, Homily for the Jubilee of the Agricultural World (2000).

26. Bacon, *Novum Organum, Book I*, Aphorisms 18–26, 44, transl. P. Urbach and J. Gibson (Chicago: Open Court, 1994). See also Harrison, "Subduing the Earth."

27. D. B. Kraybill, *The Riddle of Amish Culture* (Baltimore: Johns Hopkins University Press 1989), pp. 141–87.

28. Genetically Altered Crop Gets Test in Amish Country, Associated Press, April 30, 2001.

29. Bruce and Bruce, *Engineering Genesis*, pp. 89–92.

30. European Commission, *Communication from the Commission on the Precautionary Principle*, COM (2000) 1 final, European Commission, Brussels (2000).

31. United Church of Canada, *United Church of Canada Social Policy Positions—Genetically Modified Foods, General Principles*, 38th General Council Record of Proceedings (2003), p. 232.

32. Church of Scotland, *Genetically Modified Food*, Reports to the Church of Scotland General Assembly 1999, pp. 20/4 and 20/93–20/103 (Edinburgh: Church of Scotland, 1999). See also Church of Scotland, *Should We Become a GM Nation?* Reports to the Church of Scotland General Assembly 2004, pp. 20/4 and 20/64–20/68 (Edinburgh: Church of Scotland, 2004).

33. R. Lesseps and P. Henriot, *Church's Social Teaching and the Ethics of GMOs* (Lusaka, Zambia: Jesuit Centre for Theological Reflection, 2003).

34. Deane-Drummond, *Aquinas*.

35. U.S. Embassy to the Holy See (2004), *Feeding a Hungry World: The Moral Imperative of Biotechnology*, Conference papers, September 24, 2004, The Pontifical Gregorian University, Rome, http://vatican.usembassy.gov/policy/topics/biotech/biotechnology.pdf.

36. S. McDonagh, *So Shall We Reap*, cited in *Justice Peace and Integrity of Creation E-Newsletter*, Society of St Columba, vol. 1, Issue 12, 30 September 2004. http://www.columban.org/jpic/newsletter/september_30/documents/Critique_McDonagh.doc.

37. Bruce and Horrocks, *Modifying Creation*, pp.76–79.

38. Lesseps and Henriot, *Church's Social Teaching and the Ethics of GMOs*.

39. C. Ponting, *A Green History of the World* (London: Stevenson, 1991).

40. Society Religion and Technology Project, *Should We Clone Animals?* Information sheet (Edinburgh: SRT Project, 2004).

41. For discussion of Church views on intellectual property issues in biotechnology, see Conference of European Churches, *Critique of the Draft EC Patenting Directive* (Geneva: Church and Society Commission of the Conference of Euro-

pean Churches, 1996) and subsequent documents on website http://www.cec-kek. org/English/cs.htm; Canadian Council of Churches, *Life—Patenting Pending* (Toronto: Canadian Council of Churches, 2003); Church of Scotland, *Ethical Concerns about Patenting in Relation to Living Organisms*, Reports of the Church of Scotland General Assembly, May 22, 1997 (Edinburgh: Church of Scotland: 1997).

6

Genetically Modified Foods and Muslim Ethics

Ebrahim Moosa

Technological advances in the mass manufacture of food have radically altered the content of what we consume. Revolutionary innovations in genetic engineering and the decoding of the human genome now make it possible for vegetables in our food chain to bear animal transgenes. A tomato containing a gene harvested from a flounder may not generate repugnance in an observant Muslim, since fish is permissible for adherents of this tradition, but a potato with a pig gene may well trigger visceral repugnance. The reason is obvious: an observant Muslim is prohibited to consume pork or any of its products.

Are these developments in science and technology a cause for alarm? Do these changes impact the way we imagine the role of food in religious and ethical life? Indeed, in many religious traditions, Islam included, dietary restrictions and food consumption are constitutive of religious observance as well as the formation of identity. To what extent does technology-driven food production, especially genetically modified (GM) foods, as well as patterns of consumption of such modified foods, challenge traditional Muslim notions of food as well as the accompanying considerations of spiritual and bodily well-being?

Approach to Food

Muslim thinkers and representatives of the early tradition view food and gastronomic consumption as one aspect of the commandment to "live a good life." Addressing believers across the aeons through their prophets, the

Qur'an exhorts: "O you apostles! Partake [lit. eat, *kulu*] of the good things and do righteous deeds" (23:51). To eat of that which is "wholesome" or "good" means to partake from that which in Arabic is *tayyibat* (sing. *tayyib* or *tayyiba*). *Tayyibat* also connotes things that are lawful and permissible.[1] In fact, good, wholesome, and lawful things, according to Islamic teachings, have to be enjoyed and celebrated. Philologists helpfully and interestingly point out that the elements of pleasure and delight are included within the idea of lawfulness.

Thus the crucial responsibility of caring for the body involves both an element of observing ethics and an element of pleasuring the body. Since body and self are not essentially separated, for this reason caring for the body is like caring for the self. In other words, the integrity of the self becomes manifest and takes form in the integrity of the body. Hence, there is an elaborate practice of caring for the body, starting with the nourishment of the body in Muslim practice.

The reasoning goes like this: if the body is the vehicle for the self, then it attains the same inviolability as the self. However, Muslim teachings differ as to whether the body attains sanctity on its own, irrespective of the self (deontological) or whether the body is instrumental to the needs of the self (teleological). Since the purpose of human existence in Muslim thought is related to an afterworldly salvation it is interesting to observe how some Muslim thinkers imagined the body and its needs. In order to attain salvation the body has to be disciplined through learning and practice. Since the body is the locus of such discipline, the etiquette of consuming food also acquires a certain importance in the overall scheme of salvation.

Intelligent people know that seeking salvation in the hereafter, wrote the twelfth-century polymath Abu Hamid al-Ghazali, depends on learning and practice. "Only a healthy body can provide consistency for learning and practice," noted Ghazali, adding, "serenity of the body is not possible without food and nourishment. . . . In the light of this wisdom, some pious ancestors held that 'eating is part of the teachings related to salvation or religion (*din*).' "[2]

When it comes to food in Islam, the taboos are clearly spelled out in the normative teachings of the Qur'an and the prophetic reports, and, on the rare occasion, are to be found in the decisions of Muslim jurists who would declare some species of seafood to be abominable but not prohibited. While Muslim teachings explicitly identify foods that are banned for consumption, the tradition does not provide any reasons or rationales as to why the consumption of, say, swine and wine is prohibited. Many people reach inferential conclusions to grasp the underlying reason in the commandments that sustain religious claims. Often these are ratiocinations and not necessarily compelling evidence of the effective cause for the bans.

Historians have generally claimed that pigs and certain carnivores were banned throughout the Canaanite-Aramaic world because these animals were associated as representatives of the infernal world.[3] That may well be the reason as to why herbivorous domestic animals were deemed permissible for consumption, whereas there were reservations toward scavenging, blood-consuming carnivores in both Judaism and Islam. Muslim teaching acknowledges that wine does have certain merits as a beverage, but it also goes on to explain that its demerits outweigh the merits. Disclosure of the specific demerits of wine is absent but is often divined by scholars. Some speculate that the prohibition might have something to do with the incest taboo or might contribute to degenerate conduct. Sexual violations could occur in a state of drunkenness, as some biblical accounts suggest, even though Muslim sources make no such reference.

While there is an abundance of Muslim ethical teachings related to agricultural commerce, there are relatively few guidelines with respect to food safety. During the genesis of Islam, food safety was evidently not a major concern. Scarcity of food though was a major edificatory theme in the Qur'an. And God as the sovereign in the Qur'an is the one who can withhold food through a variety of acts of nature. Elaborate descriptions of the human condition in scenes of famine, hunger, and the fear of want are scattered throughout passages of the Qur'an. Providing food to the poor and the indigent is also seen as one of the greatest acts of kindness and compassion as well as making one worth earning divine favor.

If anything, food is viewed as a source of divine beneficence and mercy. The devout, in turn, express their gratitude to God by upholding the dignity of food and showing respect for the sources and means of food. Whether it is to safeguard food sources or for aesthetic purposes, perhaps for both sets of reasons, Muslim teachings strongly advocate the preservation of the natural environment. The protection of trees, agricultural land, and sources of water are all vital components to a balanced Muslim environmental ethics.

One overriding and forceful normative trope in Muslim ethics is the preservation of naturalness (fitra). To preserve uncontaminated nature coupled with the celebration of a healthy or sound human nature is viewed as the highest of Islamic ideals. These multiple meanings are captured by the notion of 'fitra,' a concept central to Muslim ethics and spirituality. Fitra is the inborn, intuitive ability to discern between right and wrong. "So direct your face steadfastly to faith," instructs the Qur'an, "as God-given nature (fitra), according to which God created humanity: there is no altering the creation of God" (30:30). God in the Qur'an is also described as the Original Creator (fatir), one who creates the heavens and the earth without any model.[4] The upshot is this: even nature has an ingrained disposition that determines its order.

Nevertheless, *fitra* is susceptible to distortion and corruption through sin and disobedience. It goes back to the standoff between God and Satan in prelapsarian time. Satan on refusing to bow to Adam earned divine wrath. Then Satan asked for respite to tempt the offspring of Adam, saying in the words of the Qur'an: "And I will order them to alter the creation of God" (Q 4: 119). Even though altering the creation of God here means that humans will corrupt their inner natures through disobedience and rebellion, this passage lends itself to multiple readings. While most classical exegetes have held to a metaphorical meaning, more contemporary lay and specialist readings of these verses claim that any nonremedial physical alteration to the human body or nature amounts to one acting on a Satanic inspiration.

This ambivalence toward tampering with physical nature has some vestigial relations with tradition. There is a celebrated account involving the cross-pollination or grafting of date-palm seedlings during the prophetic era of Islam. This tradition is extremely illuminating for the light it casts on two matters: first, the limits of prophetic authority, and second, on matters related to food production. The story goes that during his career in Madina, Muhammad stumbled upon farmers grafting different species of date-palm seedlings. They did so in order to produce higher yield and resistant date crops. On observing their practice, Muhammad inquired why they did so. To which they replied: "[W]e always did it this way."[5] The Prophet then retorted: "Well, perhaps, it would be better if you did not."

After some years the date harvest turned out to be a failure. The reason was attributed to the change in their agricultural methods. The dates produced were possibly few and of a very low quality, a catastrophe in a culture where dates were a vital staple. When Muhammad inquired as to the cause for such a decline in the harvest, his companions informed him that they had acted on his counsel not to do grafting. Somewhat surprised and puzzled that his companions adopted his personal opinion, he tried to rectify matters by providing some perspective as to the limits of his authority. In an often-quoted phrase, he said: "I am only a mortal. If I command you in matters of *din* (matters of moral and salvific consequence), hold on to it steadfastly. And, when I command you with my opinion, then I am a mere mortal." In another version of the same tradition, he reportedly said: "You are more knowledgeable in the affairs of your world."

There are several ways of parsing the intertextual hermeneutic at work here. One is the commonsense conclusion that Muhammad acknowledged that his dislike for grafting seedlings was based on his personal taste without any divine disfavor being communicated. Thus, when it came to agricultural matters, he conceded to experience and, in fact, endorsed expert opinion. His expertise is restricted to moral matters that explicitly impact on one's salvation. In other words, moral and ethical practices that are tied to worldly (secular) pursuits that rely on scientific or empirical knowledge are decided on their scientific and practical merits.

There is a subtle point to be made here. Human stewardship is indeed tied to salvation and how we deal with nature is part of that responsibility. But the critical question is whether matters of a secular and scientific nature can and ought to be primarily decided by teachings and inspiration that are derived from revelation. Or are these primarily matters of secular knowledge where religious stewardship and ethical responsibility ought to be the guidance. From the example provided, it became clear that Muhammad's followers at first misunderstood him. They unquestioningly acceded to his prophetic authority with resultant consequences. He later admonished that not all his actions or teachings are actuated by divine inspiration and that his followers ought to be guided by their own lights in order to grasp the human-divine fault line running through much of prophetic teachings.

To understand this, one has to turn to the notion of *din* in Islam. This term is often misleadingly translated as "faith" or "religion," whereas it has a much broader semantic connotation in Muslim discourse. In Muslim life, proper performance in the secular domain, where one strives to serve both God and humans at once, secures for the actor salvation in the afterlife. Hence, the term *din* denotes practices with salvific ends in mind, but that does not mean it excludes worldly or secular concerns. What is characteristic in Islam is that the boundary between what is typically called "religious" in post-Enlightenment conceptions of religion and the "secular" is not as clearly defined as in other Western religious traditions. In some versions of say Christianity, the "secular" is normatively separate from the "religious." Salvation in Islam depends as much on ethical performance in the work place, as it relies on devotions in the sanctified sacred space designated for worship.

But in Muslim discourse, there is an interpretative metric that evaluates commandments that have a direct bearing on the ultimate questions that is regarded in the provenance of *din*. All other matters ignored by revelation or unspoken, whether these are indirectly related to the ultimate moral values in life or are instrumental to the questions of *din*, are destined to the secular realm, *dunya*. Ethical conduct in the secular realm is assured through stewardship (*khilafa*), which is epitomized by responsibility. People can differ in their understanding as to what good stewardship means. Muhammad's own personal dislike for grafting is one thing but not good enough for public advocacy; his companions also understand him differently at times since there are other instances when they would question his rationales for prohibiting a practice or solicit reasons as to why he took a particular decision.

Careful interpreters of the Muslim tradition were acutely aware of this complexity and often divined the limits of prophetic authority. Hence, very few traditional Muslim interpreters would adduce the report of Muhammad's dislike for grafting as a normative teaching, being an analogy to demonstrate ethical disdain for genetically modified organisms (GMOs) and genetic engineering for that matter. What GMOs present are a whole series of challenges about which Muslim ethics is as ambivalent and undecided as other religious traditions. At best, Muslim ethics on a range of bioethical and scientific challenges can be described as a work in progress. Most surprising is the dearth of Muslim ethical deliberations on this topic. Apart from a few juridical responses (*fatwas*) that will be analyzed later in the chapter, even in a modern Muslim medical ethics encyclopedia there was no discussion of GMOs and safety in the entry on food.[6]

In part the muted response could in part be attributed to the daunting challenge that GMOs pose since many also view them as a benign technological advance. GMOs radically change all the inherited presumptions of a religious tradition like Islam, and present us with a dynamic system of nature, one that is a constantly emerging novelty.[7] In fact, Amos Funkenstein's observations with respect to Christian theology could easily also apply to Muslim juridical ethics and theology. Funkenstein states: "The new sciences and scholarship . . . made the traditional modes of theologizing obsolete. . . . Never before or after were science, philosophy, and theology seen as almost one and the same occupation."[8] The radical shifts in our experiences and understandings of some of the most important presumptions about nature, creation, humanity, society, the individual, and community, among other things, present us with the most profound challenges in trying to make sense of canonical opinions and traditions. In the past, the fundamental categories that we today deploy must have meant something very different, perhaps even unfathomable to us. But there is also a great deal of continuity in the meanings between present and past. Part of the hermeneutical challenge is to properly account for temporal differences in inherited teachings and make sense of them in the present.

Methodological Issues and Theory

There are a range of Muslim opinions and institutions that articulate views on ethics. Each brings a different approach, methodology, and history to its deliberations, and each vies for public attention. GMOs present particular challenges. There is nothing in Islamic teachings that designates foods to be good or bad, permissible or impermissible primarily on the grounds of some inherent quality of goodness or detestability in the food. As GMOs are unprecedented most Muslim ethicists view them as a matter that requires *de novo* personal commitment, study, and intellectual effort, designated as *ijtihad*. Over time, accumulated intellectual effort will also update the ethical canon. Muslims refer to this body of knowledge as *shari'a*, revealed teachings that are discovered discursively and often times described as Islamic law. Therefore, *shari'a* would more accurately translate as a hyphenated juridical-ethics. An interesting insight is offered by the Andalusian juridical-ethicist Abu Ishaq al-Shatibi (d. 790AH/1387AD), who claimed that "all of the shari'a is about interiorizing the most noble of character."[9] In other words, at its base *shari'a* is about virtue ethics, a fact well attested to by even some modern Muslim ethicists.

Two requirements are indispensable to the activity of *ijtihad*: The first is to grasp or comprehend the purposes of the *shari'a* or, to use a more contemporary idiom, to grasp the policy goals (*maqasid*) of the *shari'a*. The second is to acquire the skills necessary to extrapolate rules and values premised on such understanding.[10]

A major development in contemporary Muslim juridical-ethics is the shift towards utilitarianism. One clue for this is the growing reliance on the doctrine of *maslaha*, public interest or common good to resolve most ethical deliberations. In doing so, ethicists rely on the doctrine of public interest (*maslaha*) as set out by one of its twelfth-century advocates, Abu Hamid al-Ghazali. Classical Muslim ethicists have derived from their reading of the *shari'a* five positive goals or purposes. In this view all ethical and juridical practices must at all times advance and preserve religion, life, sanity, lineage (family), and property. Needless to say, the meanings of each of these goals converge with the idea of public interest. Together public interest claims provide a blanket that conceals a number of subjective predispositions held by actors that are difficult to capture in even the most intricate and comprehensive theory. A determination, for example, as to what is "property" and how it is to be preserved depends entirely on how an ethicist views the market, how he or she views the world. Does the state or do corporations have a role in regulating the market? If so, whose interests will the market serve? The moral force of the market will also depend on the economic and political ideology espoused by the ethicist: is she an economic libertarian, capitalist, or socialist? These are all very pertinent questions in deciding how one determines what is to be in the public interest.

Classical advocates of utilitarian ethics required the doctrine of public interest to meet a three-point test: the test of necessity, decisiveness, and universality. Once these tests were applied, classical theorists arrived at a hierarchical scale of benefits or interests in order of priority to list them as essential (*daruri*), supplementary (*haji*), and cosmetic (*tahsini*). Essential benefits or interests must satisfy one of the five policy goals of the *shari'a*, while the other, lower categories such as supplementary and cosmetic fulfill other, lesser goals. For reasons of fluidity and indeterminacy these goals are not explicitly stipulated and are seldom vigorously debated. Contractualist models of Muslim ethics, where moral obligation is based on fidelity to a canonical tradition of ethics with all its presumptions of governance, order, and authority reasonably understood and agreed to by all the actors, have gradually given way to utilitarian models that are often scripture-centered.[11] It is within these ethical parameters and practices that a topic such as GMOs is negotiated within Muslim ethics.

In many Muslim parts of the developing world the ethical question as to the use of GMOs first landed on the table of governments and state bureaucracies long before they become part of a process of social and ethical debate. Given the perilous state of consultative governance in many instances, debates pertaining to GMOs or cloning, to cite two examples, are not subject to deliberative ethical discussions. These issues only arrive at the desks of ethicists, more often than not, for the sole purpose that they require rubber-stamping and legitimating from religious elites who merely endorse policies adopted by their states.

Within the broad and diverse constituencies of Islamdom one can observe at least two trends on the issue of GMOs. The first are views held by a variety of traditional religious authorities as well as some technocratic and professional Muslims who support ethical and legal sanctity in support of GMOs and view the phenomenon as a manageable risk.[12] The second trend consists mainly of Muslim professionals and technocrats. This group debates GMOs in terms of a precautionary frame.

GMOs as Manageable Risk

Few traditional Muslim authorities have issued rulings on GMOs compared to the relatively more advanced discussions on genetic engineering and human gene therapy. The Saudi-based Council for Islamic Jurisprudence (CIJ), affiliated to the World Muslim League, has deliberated on GMOs since around 1998. The question about GMOs was put indirectly to the CIJ, in the context of providing general guidelines on genetic engineering.[13] In its ruling the CIJ permitted the use of genetic engineering for therapeutic and preventative purposes, provided that its use did not produce a greater harm than the impairment it attempted to remedy.

For guidelines more pertinent to genetic engineering and food production in the CIJ wrote: "It is permissible to employ genetic engineering and its attendant products in the sphere of agriculture and animal husbandry. This is allowed on condition that all necessary precautions be adopted in order to prevent any kind of harm—even on a long-term basis—to humans, animals and the environment."[14] The CIJ's directive on genetic engineering and food is conspicuous for its brevity and nebulous language. While the CIJ permitted the use of genetic engineering in the food chain, it issued a vague rider cautioning against harm in its use. It gave little guidance as to how harm is calculated. Who decides on the level and extent of harm? It becomes evident that the CIJ directive had no independent viewpoint on the issue of GMOs or on the larger discussion on genetic engineering for that matter. On all these issues it relied almost entirely on expert knowledge, scientists, and technologists who inform the CIJ and on the basis of which they formulate the ethical permission to use GMOs coupled with the caution against harm. On what grounds the CIJ permits the use of GMOs remains unclear. Is it because they are unconvinced that the use of GMO crops will decrease or disrupt gene variation? Or do they favor the viewpoint that GMOs help with nutritional deficiencies and aid in producing higher yield crops? Since the reasons and rationales for favoring GMOs are not given, therefore the favorable decision of the CIJ signed by some leading jurists from the Muslim world remains under a cloud.[15]

However, to be fair, on face value it appears that CIJ did show some awareness of the gravity of genetically engineered substances in the human food chain. For in the very same fatwa, it insists that the use of GMOs in food and medicinal products be disclosed through labeling. Consumers wary of potential harm from GMOs have a right to be informed. "The Council calls on all companies and manufacturers of foods and medicines and other products derived from substances produced by way of genetic engineering," the fatwa read, "to disclose the contents of such [engineered] substances in order to ensure transparency as well as to alert users to [possible] harm and [inform them about products] that are prohibited in terms of juridical-ethics (shar')."[16] If harm is anticipated in the long term to humans, animals, and the environment, then the CIJ finds this to be compelling grounds to apply the rule of precaution. Due to the absence of detailed documentation it is difficult to grasp how the CIJ debated the political, economic, and other scientific issues associated with GMOs. These aspects cannot be easily delinked from the strict ethical and moral concerns related to the ingestion of foods that may potentially have harmful consequences.

What generates considerable concern, if not visceral disapproval and skepticism, among observant and devout persons is the use of genes derived from prohibited substances. Once again manufacturers are required to inform consumers when such substances and transgenes are used in food and medicines. So the question arises: would a potato developed with a pig gene, or a medicine using heparin, a substance derived from pigs and other animals, be permissible for Muslim consumption? One of the methods applied by traditional Muslim jurists was to inquire whether the essential physical and chemical properties of a prohibited substance had been "transformed" (*istihala*) into a totally different product. If so, then the original ban on the prohibited substance no longer applied.[17] Since the prohibited substance had mutated into something else, like wine mutates into vinegar, the rule no longer applies.

If the rule of the transformation of a prohibited product is applied to transgenes it could have interesting outcomes. Some jurists have been applying this analogy to their views on transgenes. The question would be: does the transgene retain any of the essential features of the prohibited substance, or does the gene merely serve as a chemical formulation outside the potato to produce another substance, which is then in turn introduced into the potato? In fact, one can even raise the question whether one can really say that a pig *gene* is in the same degree of prohibition as pig *meat* itself. Some would argue that what is prohibited for Muslims is the consumption of pork. Can pork be viewed to be identical to pig genetic materials, which "code for the nucleotide sequences in mRNA, tRNA, and rRNA required for protein synthesis"?[18] Obviously, opinions would be divided on these questions depending on a range of factors. Among these factors would be the way adherents understand the moral commandments of their faith, their meritorious view of science or otherwise, and the way in which certain realities get communicated to them. Some scholars might dismiss any similarity between pig genes and pork, while others would view the two to be identical in essence. These issues have been discussed somewhat extensively in certain traditions of Jewish ethics. One of the concerns among Jews was whether the insertion of a gene in a tomato produces phenotypic change. The phenotype of an individual organism is either its total physical appearance and constitution or a specific manifestation of a trait, such as size, eye color, or behavior that varies between individuals.[19] Some in Jewish circles object if transgenes produce genotypic changes, in other words, changes to the specific genetic makeup of an individual in the form of DNA. Muslim ethicists would most likely follow a similar line of inquiry and ask similar questions.

Following in the footsteps of the CIJ, Muslim religious authorities in Indonesia, which hosts the largest Muslim population in the world, have approved the use of genetically modified organisms in the human food chain. Singapore's Muslim religious authorities have also endorsed the use of GMOs, as have those in Malaysia. In India some Muslim religious authorities, generally more conservative compared to their colleagues in the Middle East, have given a cautionary green light to GMOs. As far back as October 1997 at its meeting in Bombay, the Islamic Fiqh Academy of India adopted a resolution approving the cloning of animals and plants. It qualified its approval by saying it deems permissible "such cloning in plants and animals that are beneficial to humans and which do not threaten humans from a religious, moral and physical perspective."[20] This openness to the altruism of science is premised on thin versions of Muslim ethical doctrines and has emboldened other institutions to follow suit. In 2005, Darul Uloom Nadwatul Ulama in Lucknow, India, an important seminar on the subcontinent, issued a fatwa approving GMOs. Asked whether the consumption of GMOs was permissible, the brief reply issued by both Mufti Masood Hasan Hasni and Mufti Nasir Ali read: "To eat such fruit and vegetables is permissible. Unless the harm of a thing is known categorically or by means of a dominant probability, one cannot designate a permissible thing to be prohibited on a mere apprehension of harm. However, if out of precaution one refuses to partake of such foods, then that is the exercise of one's choice.[21]

In Malaysia, Abu Bakar Abdul Majeed sees absolutely no conflict in the goals of Muslim ethics and the use of GMOs as long as the dietary requirements of Islamic law are observed.[22] Transgenes derived from non-permissible food sources would, in his view and those of others, constitute such a prohibition. Majeed fosters the view that economic gain and the advancement of the standards of nutrition are both highly desirable goals. If biotechnology served such ends within an acceptable Islamic ethical paradigm, then it is permissible. In his view the ends justify the means. Majeed draws on the work of bioethicist David Magnus who disapproves of reducing the question of biotechnology and GMOs to a risk-benefit analysis. For Magnus, as well as Majeed who follows him, the following question arises: is it so morally wrong to recombine genes from different organisms? Of course, they answer their question rhetorically. As much as Majeed is enthusiastic about genetically modified foods, he never directly addresses the moral question whether it is permissible to recombine genes. Neither does he provide any theological or ethical backing for his argument. From his comments it becomes clear that his approval of GMOs is informed by a highly pragmatic approach to science.

Two individual scholars shed some realist light on the debate on GMOs. A noted traditional scholar from Egypt, Shaykh Dr Hasan 'Ali al-Shadhili, showed some sensitivity and awareness that the flourishing of the human race requires a high level of genetic diversity. While he is positive about biotechnology, he is also concerned that it can go in the opposite direction with harmful effects.[23] Iran's Ayatullah Muhammad Ali al-Taskhiri was more optimistic about the outcome of biotechnology. He hews close to the scientific debates and invests a great deal of authority in the judgment of science. Ethical and political decisions, Taskhiri points out, cannot be made in advance of the scientific evidence. "There should be no haste in making ethical decisions as long as the scientific results are inconclusive," he said.[24]

The Islamic Food and Nutrition Council of America (IFANCA), the main North American certifying body that designates food as "permissible" (halal) by Islamic standards, is reported to support foods derived from

GMOs.[25] However, IFANCA has not issued an explicit statement in support of GMOs. In one newsletter IFANCA said: "On the issue of biotechnology and GMO foods, there is no specific reference to this in the Islamic Law sources. However, it is accepted that GMO from *haram* [prohibited] sources would be *haram* [prohibited]. There is a place for biotechnology in *halal* food production. One example is the production of GMO chymosin for cheese manufacture."[26] Other than this statement, IFANCA appears ambivalent and demonstrates some concern about GMOs, animal feed, and food safety. "Islam stresses the need to consume 'pure' foods," IFANCA stated. "Organic products are a major step in the direction of halal [permissible food]." Then IFANCA makes a surprising theological leap. It points out that since God is both the author of the laws prescribing permissible foods (*halal*) and the creator of humans, the pietism that God "knows what is best" for humans must be true.[27]

However, Richard Foltz has criticized IFANCA for its uncritical support for modern food production techniques.[28] IFANCA followed the letter of law, Foltz said, and as a result mechanically applied the rules to very different kinds of foods and food production techniques. Foltz proposed that it would be more profitable if IFANCA and others tried to grapple with the underlying spirit (health, compassion) of Islamic law on food, which might produce a different set of answers. As a certifying council, IFANCA restricts itself to meticulous investigation of food contamination with pork and alcohol products but paradoxically embraces GMOs that might incubate harm of an unknown kind.

Precautionary View

In Muslim communities, especially those in the West, religious authorities are not among those who exhibit greater sensitivity and concern about GMOs in the nutritional system. Rather, modern-educated persons with scientific training and backgrounds are among those who harbor the strongest reservations, with the help of rudimentary elements of a scriptural Islamic theology that is further informed and amplified by a critical and self-reflexive view of science. Among those who adopt such a theology is Mohammad Aslam Parvaiz, who unhesitatingly aligns himself with that sector of the scientific community who believe that the use of transgenes in food harbors catastrophic consequences.[29] Parvaiz catalogues an array of scientific opinion and studies to show how transgenic fish and crops already pose serious environmental risks.

However, it is not always clear whether someone like Parvaiz is primarily motivated by scientific convictions or by scriptural theology or a combination of both. His theological starting point is that nature had been designed or had evolved to reach a delicate and balanced (*mizan*) ecosphere. *Balance* is a term derived from the Qur'an that imports the image of the weigh scale of justice and fair balance in life. Within Muslim thought this notion has been elevated to a metaphysical principle of proportionality and measure (Q 54:49,55:7–9). Parvaiz views genetic engineering as janus-faced. While it produces "many innovations in Allah's creations," the very same technology is the source of deleterious "change [in] Allah's creations." His principal concern is that genetic engineering manipulates microorganisms about which we know very little. Transgenic organisms irreversibly interfere with the "most sacred of sacreds, that is, the gene pool of an organism," Parvaiz argues. Here he is most probably referring to genotypes and expresses doubts whether any scientific grounds can justify altering certain life forms to suit human needs. Furthermore, his major concern is that such an invasive level of human interference in the ecosystem will produce unforeseen disturbances on the planet. Parvaiz, in his resistance to GMOs, resorts to a theological reading of certain passages of the Qur'an, especially the verses that urge humans not to alter God's creation.

Parvaiz's interpretation of the Qur'an is indeed very modern. Classical Muslim exegetes did not understand the ban on the altering of nature to mean a prohibition on manipulating physical nature for meaningful ends. Rather, they believed that it was human nature that was susceptible to distortion, which could in turn result in morally degrading practices. Here lies the crucial difference between those traditionalists who support the manageable risk position on GMOs and some modernists who adopt the precautionary approach. Some of the Muslim traditionalists clearly understand the Qur'anic proscription to apply to our moral nature, not our physical nature. The Muslim modernists, in turn, read those same passages and interpret their meanings to be a prohibition on the corruption of physical nature. While the differences in practice might be subtle, they hold serious consequences for theology. Indeed, a theological ban on the manipulation of nature often produces a certain kind of absolutism and determinism. If the traditionalists are pragmatically restrained by the canon, then the modernist exuberance with science can potentially result in the rendering of absolutist interpretations of the Qur'an.

In a less theological mood, Saeed A. Khan urges Muslim communities to join the alliance of concerned scientists, producers, and consumers in the United States and abroad to combat the use of GMOs. Such opponents view GMOs as an alarming and illicit use of science that has the potential to "colonize" not only people residing in the developing world but also people living in the developed West through scientific and corporate hegemony.[30] Khan draws our attention to global policy concerns. While he acknowledges that GMOs could be of benefit, he is not altogether sure if the benefits-versus-dangers algorithm has been satisfactorily addressed. He is prepared to err on the side of caution, arguing that if unregulated and unchecked then food from GMOs has the potential of "becoming a weapon of mass destruction, rather than the intended weapon of mass consumption."

The 1997 Casablanca resolution of the Ninth Islamic Medical Ethics conference serves as a capstone for this cautionary approach. "It will be excessive to claim that genetic engineering was safe in the realm of plants, despite many years of experimentation," the resolution stated. "And thus far, it is not even on the threshold in terms of its application to animals. Perhaps the unknown factor is the greatest cause for anxiety in this matter. Humanity should certainly not forget the lesson learned most recently from the splitting of the atom: the detrimental consequences of radiation to the body became apparent, a fact that was neither known nor expected at the time of the invention. Therefore, it is imperative to monitor the outcome of genetic engineering experiments on plants and animals for a protracted time."[31]

Views from the Focus Group

Viewpoints from both the managed risk approach and the precautionary approach percolated the opinions canvassed at the Muslim focus group held in Vancouver, BC, on the question of the relationship of religiously informed dietary practices with food.[32] The focus group elicited a series of very productive and interesting responses as to how communities relate to food. The voices of ordinary Muslims in ethical debates are seldom heard; often the interpreters of the tradition have the final say, often without broad consultation with communities. This study is informed by the rare input of a cross-section of opinions held by Muslims and provides an index of the sensitivities and concerns they harbor with respect to food.

Several participants believed that the Qur'an encouraged Muslims to pay special attention to certain types of food. Most participants found the mention of foods in the Qur'an, such as milk, honey, figs, grapes, and pomegranates to confer a certain sacrosanct status on those food types, which made them approach these foods with a certain piety. One participant had given thought to such food types and classified it in the "encouraged" category of Islamic law. Another reason why some viewed the foods mentioned in the Qur'an differently was because these foods were "naturally originated." Others were drawn to a substance like honey, because they believed it to be particularly "nutritious," rich in vitamin K and fructose. Honey also had other properties and uses, such as the ability to combat certain allergies and heal wounds, one participant added. Not only did certain foods have nutritious properties, but they also had healing and spiritual characteristics, some participants said, singling out foods such as honey and milk.

Some members of the focus group offered medical and scientific rationales as to why Islam forbade the consumption of carrion, blood, and pork. Blood contained harmful pathogens, they claimed, which could in turn be harmful to the body. For this reason Islamic slaughtering methods of allowing the blood to drain from the meat were viewed to have health benefits. Participant V8 succinctly conveyed a prevailing attitude toward food. "The philosophy of Islam," V8 said, "is to save the human body from any dangers. So Islam prohibited . . . different food[s] [since] they are injurious to human health."[33] Pork, participant V3 claimed to learn from a son who is also a neurologist, contributed to the parasitic infection of the brain, neurocystecercosis, with symptoms similar to epilepsy.[34] The same participant explained that to observe a religious diet was to seek God's love, friendship, proximity, and wisdom. Another person, V2, provided a more sugary rationale, arguing that the dietary prescriptions of Islam dating back to antiquity contained wisdom that even science could not fathom.

Participant V7 aired greater skepticism about science and thereby its claims about food safety and GMOs. This participant claimed that "science always takes the U-turn," meaning that later findings of science could contradict or correct earlier findings. By contrast, V7 continued, "if you analyze the Qur'an, it's never taking U-turns." While the triumphant tone favoring faith-based claims is self-evident, of greater significance is the sense the interviews convey of practitioners' views that religious teachings on food contain a higher wisdom than the claims made by science about foods. However, the same participant also appealed for a "more open-minded" approach in order to shorten "the gap" between religion and science.

Participants in the focus group demonstrated a heightened awareness about the importance of labels, since there is a strong suspicion that certain substances prohibited by Islamic rules might be utilized in foods as innocuous as cakes and pastries. Many were concerned about contamination of prohibited products. "Coming to Canada," one participant said, "we always look at the labels and what's in the product, what we can buy, and what we cannot buy, and it's a challenge." Once the full extent of the debates about GMOs dawned on some of the participants, several also insisted that genetically modified foods ought to be labeled.

GMOs elicited a spectrum of opinions among the participants, in part reflecting the lack of clarity found in the meager Muslim scholarly literature on the topic.[35] Overall the focus group participants showed hesitation and ambivalence towards embracing GMOs and their use in the human food chain. Predictably, the use of a prohibited gene, like the use of a pig gene to enhance tomatoes, was met with strong disapproval, since pork was unlawful

to Muslims. Another participant, V8, who followed the Ja'fari interpretation of Islamic law adopted by those belonging to the Shia sect, objected
to consuming cucumbers that were modified with jellyfish genes, since most
seafood, apart from fish, were not allowed in Ja'fari law.

Objections to GMOs ranged from particular understandings of Islamic
law and theology to making certain presumptions about what is natural
and balanced in terms of an Islamic worldview. For V7 it was not so much
that certain genes were stained by rules of prohibition; rather it had to do
with a natural balance in nutrition that GMOs would disturb. "So when
we genetically modify a food," V7 said, "what happens is we are creating
one ingredient; we are creating an imbalance into that food, right? . . . [Y]ou
are creating an imbalance in the nutritional value of that particular food."
The main concern was about the unknown and hidden effects of GMOs.
"Since you are creating an imbalance," said V7, "it has multiple after-effects
of that food, right?"

Some participants showed an awareness of the grafting of date-palm
seedlings in prophetic times and the Prophet's later approval of the practice. So V3 was very much in favor of putting wisdom, which was seen as
a God-given endowment, to beneficial use. But the principle concern was
that such wisdom must not lead to a situation where lawful and unlawful
items were mixed. More important, V3 had strong reservations about mixing
genes from one species to another. Taking genetic materials from animals to
plants and vice versa was to V3 an instant when one disturbs the balance
of nature. Mad cow disease alleged to stem from the use of animal products
in the food chain for herbivores such as cows was offered as an example
of mixing the genetic makeup of the food in the food chain from different
species and the harm it causes.

Among the participants, V8 pleaded for greater nuance and pointed
out that there were some contradictions in the views posited by fellow
Muslims in the focus group. Gene transfers involved the moving of invisible particles, V8 said. Gene transfers of prohibited substances were different
from genetic modification involving lawful substances. Genetic modification
already occurred in nature, V8 said, and human beings only accelerated the
modification process via technology. If a gene from a pig were to be used
to prevent the transfer or spread of mad cow disease, then V8 would be
positively predisposed towards such modification and use. Another participant, V1, was also favorably predisposed to genetic modification, arguing
that Islam was not opposed to such practice. These participants detected
a weakness and lack of complexity in the reasoning adopted by other participants in their various attempts to address an issue as serious as GMOs.
They wished to adopt some distance from an oft-displayed religious reflex
that caused many adherents as well as traditional experts to take refuge in

the binary permissible (*halal*) or impermissible (*haram*) logic of Islamic law. As to the use of prohibited substances in GMOs, V1 pointed out that the use of "one gene is negligible." However, like V8, V1 also preferred that Muslim experts be the final arbiters of this question. They both expected experts in religion to be in conversation with scientists in order to decide on such weighty matters as GMOs.

Participant V5 offered some advice to Muslim bioethicists, especially when deliberating on GMOs. One should determine this issue not only on the grounds of permissible or impermissible substances, V5 said. "I think Islam . . . looks from many other perspectives also before it can say whether something is acceptable or not acceptable." Of specific concern to V5 was this:

> [W]hat would be the effect of [GMOs] on the environment short-term and long-term? What would be the effect of [GMOs] on the health of the people who consume [them]? What would be the effect of [GMOs] on the ecosystem, right? . . . I have a verse from the Qur'an where it says that there is no changing in God's creation. And, in another place the Qur'an talks about the balance which brother V7 mentioned. I think the balance is not only nutritious balance . . . it's talking about balance in the absolute sense. . . . So if we modify something and it has an effect which is not wanted, how do we take care of that? . . . I don't think Islam can say much on this point at this time because the results are not in front of us.

This participant demonstrated a sophisticated understanding of "balance" as not merely being a matter of metrics, but that it involved something more subtle. One is loath to second-guess the intentions of focus group participants, but here there is more than a hint of maintaining a metaphysical and cosmic balance, which includes the nourishment of the human body as well as the maintenance of the ecosystem.

Several of the participants were dubious of the long-term effect of GMOs and urged caution and appealed for a slow-down in its application. They feared that technology was hurtling forward at breakneck speed without any pause, a prospect that might lead to a situation of no return with unquantifiable possibilities of harm, especially if GMOs swamped the food chain. Others canvassed expressed a certain helplessness and incapacity to deal with issues of food at this level of complexity. At the same time they also displayed a certain pragmatism mingled with anxiety.

In many ways the reasoning of some of the participants in the focus group resembled the theology utilized by those Muslims who had specialized in science and who held firm views against the use of GMOs. Many of the laypersons participating in the focus group inflected the idea of human stewardship (*khilafa*) and responsibility in human relations with the ecosystem. Surprisingly, this overarching ethical theme was absent in the Muslim scholarly literature produced by the traditionalists.

GMOs and Muslim Ethics

Not only GMOs but also a whole range of modern practices in bioethics ranging from organ transplantation, brain death, gene therapy, and now genetically modified foods pose challenges to the inherited legacy of Muslim ethics. As true as Funkenstein's insight shows how religion and science have become delinked in the modern period, there are other more complex challenges.

Muslim ethicists, whether they are of the traditional stripe or of the modern scientifically trained kind, have yet to configure the theory and practice of Muslim ethics in relation to vastly changed social realities. It is self-evident from the snapshot of views on GMOs that the mechanistic application of either traditional rules of the *shari'a* or enlightened interpretations of scriptural authority at best produces a hit and miss version of ethics, and at worst is deeply unsatisfactory. In part, this has to do with the intellectual lay of the land in ethical thinking in different parts of the Muslim world and in places where significant communities of Muslims find themselves.

The critical fault line lies with those who advocate a normative ethics, who determine, judge, and justify practices with a tool kit of knowledge from the canon. This canon has been slow to update itself and carries a baggage of norms, practices, and forms of reasoning that no longer resonate with contemporary experiences of Muslims. Hence, anachronism is often the product of such reasoning, even though such traditional modes of thinking can be utilized for pragmatic ends to justify modern practices. But whether this approach restores intellectual integrity remains questionable. It therefore comes as no surprise that the rulings from traditionalist practitioners, such as the CIJ and the Nadwatul Ulama, can with unmitigated clarity rule on the permissibility of GMOs with a perfunctory paean to caution.

Even those who advocate a precautionary approach to GMOs provide a rather thin theology to justify some emotional resistance to the use of GMOs. In fact, the case of GMOs and food safety is not really about whether a food product is lawful or unlawful. Rather, the desirability of GM food products actually forces us to think about larger questions, such as what are desirable lifestyles and life forms? How do we utilize resources on earth? In other words, meaningful ethical discussions on GMOs ought to reflect fairly serious thinking about how individuals and communities envisage what they deem to be desirable lifestyles, practices, and truth claims. Unless one has thought about some of these metaissues and adopted a political position for or against certain decided and desirable ends, only then can ethical judgments for or against GMOs cohere and make sense.

Ethics is a discourse, a system of interwoven truth claims that are embedded in social relations and material practices. Since it is embedded in social relations and material practices it is also in a certain state of flux and part of a network of relations. Things converge and diverge from old and new positions. New elements merge with old ones to become starting points for new modes of thinking in keeping with changing experiences and newer forms of knowledge. In order to deal with a range of contemporary ethical issues, I recommend a constructivist approach to ethics, one that indicates a particular way of understanding the relation between knowledge and what we experience as reality.[36] In constructivist accounts, the way we perceive the world in terms of truth, cognition, science, and related matters is not independent of sensory and conceptual-discursive activities, but rather, as Barbara Hernstein Smith put it, our knowledge emerges from or is constructed by those activities and experiences.

Furthermore, a constructivist ethics must take cognizance of the fact that in a range of issues in contemporary life, but especially with respect to the industrial reshaping of life and the turn to science, we enter into a very different kind of discursive struggle that fits with what Michel Foucault had termed "biopolitics."[37] Foucault helps us to understand what modernity changed in terms of the way we perceive ourselves and our specific notions of existence. It is our existence as social and living beings that modernity questions. This is the new "politics," as Peter Andrée explains, in the sense that the ways we come to understand life are increasingly constituted through scientific truths. And these truths are more than "facts" since they are value-laden in their framing. Thus the biopolitics of genetic engineering and the way it plays out in concerns about food safety and GMOs are not merely decisions taken about "facts," but they center around certain sets of values.

The significant point Foucault makes is to claim that unlike ancient governance, where sovereign power was marked by the right to take a life, modern power became obsessed with the "administration of bodies and the calculated management of life."[38] Biopower today has its parallel in ecopower or geopower with its distinct object as a concern with "life" accompanied by a wide array of knowledges and practices.[39] Power no longer deals with modern human beings as legal subjects but deals with living beings and seeks to attain mastery over them. GMOs fall squarely within the flux of contemporary iterations of biopower, namely, ecopower. Given the dynamic nature of biotechnology, neither rejectionist nor panglossian approaches result in productive engagements with biopower. Rather, a discursive approach allows one to see both the productive and disciplinary character of ecopower and especially the numerous points of resistance that exist within discourses.

Beyond a simplistic scripturalism and a decontextualized tradition-alism, a constructivist ethical turn will enable Muslim ethics to address the challenges of a radically changing ecosphere with greater conviction and efficacy. Provisionally, my own position is molded in the precautionary frame toward GMOs for a number of reasons. First, there are compelling reasons why one should be cautious, given the fact that risk management is no guarantee against irreversibly harming the ecosystem. Second, both the precautionary and risk-management frames acknowledge the links among norms, science, and risk in decision making. In that sense they are opposite ends of the same spectrum. Third, one needs to explore GMOs and food regulation within a framework that is not restricted by science-based decision making. Unfortunately, religious communities, including Muslims, have not been able to go beyond science-based decision making on this issue.

Conclusion

Lack of vigilance against GMOs and the inability to probe beyond science-based decision making also has to do with the location of Muslim communi-ties. Many live in the developing world where food scarcity, international aid, and food imports are subject to international and geopolitical consid-erations. Political and economic exigencies may prevent religious authori-ties from investigating biotechnology with a hermeneutics of suspicion. To do so would require religious authorities to be prepared to politically and intellectually challenge political authorities, which in many places could be a risky and intimidating venture. Such an approach could come with attendant political risks, and the intellectual skills and resources to counter the hegemonic scientific discourses would often be lacking.

Science and technology are often presented as a panacea for the prob-lems of the developing world with promises of life-changing possibilities and the potential to save lives. In the case of vaccinations and other forms of health care, this is not a totally untrue proposition. With science's prophetic aura increasingly becoming a marker of "civilization" few people have the power and authority to resist the darker side effects of this juggernaut. Hence people, including religious authorities, come under the ideological spell of science and do not always possess the skills, moral as well as critical, to resist and question science and its practices. The absence of democratic governance compounds the moral issues. Often religious authorities merely have to endorse government decrees instead of facilitating public debate.

Religious authorities take a very pragmatic approach in evaluating biotechnology associated with GMOs. Since there is no compelling evidence that GMOs can harm the body, coupled with the fact that it is rare that prohibited transgenes are used in food, Muslim religious authorities are willing to give GMO-producing biotechnology the green light. Whether the long-term environmental impact of frontier biotechnology will be subject to serious ethical and moral scrutiny within Muslim quarters is not always evident. However, there is some hope that those who adopt a precautionary approach might be instrumental in expanding the parameters of the debate and thereby bring a larger set of issues and concerns into ethical and discursive purview.

Notes

1. Muhammad b. Mukarram Ibn Manzur, "Lisan Al-'Arab," ed. 'Abd Allah 'Ali al-Kabir (Cairo: Dar al-Ma'arif, nd), 4:2733.

2. Abu Hamid Muhammad b. Muhammad al-Ghazali, "Ihya 'Ulum Al-Din" (Beirut: Dar al-Kutub al-'Ilmiyya, 1421/2001).

3. S. Daniel Breslauer, "The Vegetarian Alternative: Biblical Adumbrations, Modern Reverberations," in *Food and Judaism*, ed. J. Leonard; Ronald A. Simpkins Greenspoon, and Gerald Shapiro (Omaha: Creighton University Press, 2005), p. 83.

4. Thomas Cleary, *The Qur'an: A New Translation* (Starlatch, 2004).

5. Muslim Ibn al-Hajjaj, ed., *Sahih Muslim*, 9 vols. (Cairo: Dar Abi Hayyan, 1415/1995), 8:128.

6. Ahmad Muhammad Kan'an, *Al-Mawsu'a Al-Tibbiya Al-Fiqhiya* (Beirut: Dar al-Nafa'is, 1420/2000).

7. Ronald Cole-Turner, "Theological Interpretations of Biotechnology: Issues and Questions," in *Claiming Power over Life: Religion and Biotechnology Policy*, ed. Mark J. Hanson (Washington DC: Georgetown University Press, 2001).

8. Amos Funkenstein, *Theology and the Scientific Imagination from the Middle Ages to the Seventeenth Century* (Princeton, NJ: Princeton University Press, 1986), p. 3.

9. Abu Ishaq Ibrahim b. Musa al-Shatibi, *Al-Muwafaqat Fi Usul Al-Shari'A*, ed. 'Abd Allah Daraz, 4 vols. (Beirut: Dar al-Ma'rifa, nd), 2:77.

10. 4:105–06. Some scholars prefer to give *ijtihad* a restricted scope, stating that juridical-ethicists can only engage in activism in the absence of clear directives from the revealed sources and the rules established by precedent. To make it even more restrictive some require contemporary experts in juridical ethics to abide by the onerous and largely dated methodology and hermeneutics established by the canonical schools of law (*madhahib*). For orthodoxies of this stripe then, adhering to

the formalistic rules makes all the difference between an acceptable and legitimate production of *ijtihad* and one that falls short of such standards that impute legitimacy or its opposite. Countering this more formalistic school are others who argue that even when clear directives, texts, and statutes do exist, ethico-judicial activism, *ijtihad*, is integral to the moral enterprise, albeit a highly circumscribed activism. For it could be argued that every act of interpretation such as choosing one precedent in favor of another or preferring one analog above another involves an element of activism. In fact, al-Shatibi shows instances where even transparent verses of the Qur'an are subject to interpretive *ijtihad* (al-Shatibi, *Al-Muwafaqat Fi Usul Al-Shari'A*, 4:103). He further claims that *ijtihad* is a requirement in every age. The reason for his claim is self-evident but often missed: textual rulings that constitute the early legal sources are based on a number of finite events. Therefore, such texts by their very nature have in-built limitations, Shatibi argues, whereas contingencies are unlimited. Therefore, the fact of temporality requires an ongoing process of *ijtihad*. (Shatibi: 4:104.11). Apart from trying to explain some of the difficulties in making ethical determinations within contemporary Muslim juridical ethics, this necessary diversion also sheds light on another dimension. Practices involving GMOs, genetics, and cloning, as well as brain death, organ transplantation, and a range of issues in biotechnology are all very much science- and technology-driven enterprises. If we keep in mind my earlier observation that the relationship or fit between inherited theories of Muslim juridical ethics and the presumptions of scientific tradition have been ruptured and that their resolution remains a work in progress, then issues related to GMOs or any of the ethical dilemmas raised by modern technology are profound questions of translation. Given the diversity of Muslim societies, subcultures, and economic and political disparities ranging from developed world to developing world contexts and a myriad of intermediate situations, it is extremely difficult to speak in magisterial terms about any of these issues without losing some of the rich textures of what ought to be discussed. One of the disadvantages of such decontextualized discussion is that the ethical deliberations are extremely mechanical in nature.

12. I am indebted to Peter Andrée, "The Biopolitics of Genetically Modified Organisms in Canada," *Journal of Canadian Studies* 37, no. 3 (2002) for the helpful categories of framing the GMOs debate in terms of manageable risk and the precautionary approach.

13. Majlis Majma' al-Fiqhi li Rabitat al-'Alam al-Islami, "Bi Sha'n Istifada Al-Muslimin Min 'Ilm Al-Handasa Al-Wirathiya" (Pertaining to Muslims Deriving Benefit from Genetic Engineering), http://www.themwl.org/fatwa/ (1419/1998).

14. Ibid.

15. The thinkers on the IJA committee included the late Shaykh Abd al-Aziz Abd Allah bin Baz of Saudi Arabia, the late Shaykh Mustafa al-Zarqa of Jordan, Shaykh Yusuf al-Qaradawi based in Doha, Qatar and the late Mawlana Abu al-Hasan Ali al-Nadwi from India, in addition to some thirteen other signatories.

16. al-Islami, "Bi Sha'n Istifada Al-Muslimin Min 'Ilm Al-Handasa Al-Wirathiya (Pertaining to Muslims Deriving Benefit from Genetic Engineering)," http://www.themwl.org/fatwa//.

17. iMajlis al-Majma' al-Fiqhi bi Rabitat al-Alam al-Islam, "Bi Sha'n Hukm Ist'Mal Al-Dawa Al-Mushtamal 'Ala Sha'y Min Najis Al-'Ayn: Ka Al-Khinzir Wa

Lahu Badil Aqall Minhu Fa'idatan Ka La-Hebarin Al-Jadid," http://www.themwl. org/fatwa// (1424/2003).

18. http://en.wikipedia.org/wiki/Genetics.

19. http://en.wikipedia.org/wiki/Phenotype.

20. Qadi Mujahidul Islam Qasmi, "Jadid Sa'insi Teknik: Cloning" (New Delhi: Islamic Fiqh Academy, India, 2000), p. 75.

21. Fatwa issued 24/08/1426 AH, Darul Ifta, Nadwatul Ulama.

22. Abu Bakar Abdul Majeed, ed., *Bioethics in the Biotechnology Culture* (Kuala Lumpur: IKIM Institute of Islamic Understanding Malaysia, 2002).

23. Hasan 'Ali al-Shadhili, "Al-Istinsakh: Haqiqatuhu, Anwa'Uhu. Hukm Kulli Naw' Fi Al-Fiqh Al-Islami," *Majallat Majma' al-Fiqh al-Islami: al-Dawra al-'Ashira* 10, no. 3 (1418/1997), p. 204.

24. Ayatullah Muhammad 'Ali al-Taskhiri, "Nazratun Fi Al-Istinsakh Wa Hukmuhu Al-Shar'i," *Majallat Majma' al-Fiqh al-Islami: al-Dawra al-'Ashira* 10, no. 3 (1418/1997), pp. 219–20.

25. Rob Lake, "Current Awareness of Genetically Modified Food Issues Project 99," http://www.Moh.Govt.Nz/Moh.Nsf/7004be0c19a98f8a4c25692e007bf833/ 4bfbc7660fe2d8d44c25695f007efbda/$File/September2001.Pdf#Search=%22ijc%20 and%20gmo%22 (Institute of Environmental Science and Research Limited & Christchurch Science Centre, 2001).

26. http://www.ifanca.org/newsletter/2003_08.htm.

27. http://www.ifanca.org/newsletter/2003_01.htm.

28. Richard C. Foltz, *Animals in Islamic Tradition and Muslim Cultures* (Oxford: Oneworld, 2006), p. 117.

29. Mohammed Aslam Parvaiz, "Scientific Innovation and Al-Mizan," in *Islam and Ecology: A Bestowed Trust*, ed. Richard C. Foltz, Frederick M. Denny, Azizan Baharuddin (Cambridge, MA: Harvard University Press, 2003).

30. Saeed A. Khan, "Neocolonialism in the Muslim World through Genetically Modified Foods: The Empire Strikes Back?" (Association of Muslim Social Scientists, 32nd Annual Conference September 26–28, 2003).

31. Islamic Fiqh Academy, "Tawsiyat: Al-Nadwa Al-Fiqhiya Al-Tibbiya Al-Tasi'a Cassablanca 1418/1997," *Majalla Majma' al-Fiqh al-Islami* 10, no. 3 (1418/1997), p. 431.

32. Muslim Focus Group, May 30, 2005, Vancouver BC.

33. Ibid., 13.

34. "Epilepsy," *Encyclopædia Britannica Online*. Accessed September 23, 2006 from http://search.eb.com/eb/article-9032798.

35. Muslim Focus Group, May 30, 2005, Vancouver, BC.

36. Here I draw upon the work of Barbara Hernstein Smith, *Scandalous Knowledge: Science, Truth and the Human* (Durham, NC: Duke University Press, 2006).

37. Michel Foucault, *The History of Sexuality Volume 1: An Introduction*, trans. Robert Hurley, 3 vols., vol. 1 (New York: Vintage (1978), 1990), p. 138–44.

38. Ibid., p. 140

39. Andrée, "The Biopolitics of Genetically Modified Organisms," p. 167.

7

A Hundred Autumns to Flourish

Hindu Attitudes to Genetically Modified Food

Vasudha Narayanan

What do a billion people belonging to various economic classes and myriad social divisions, who now live in every part of the world, think of genetically modified (GM) food in a variety of ritual, social, medical, and economic contexts? Would Hindus accept, reject, or joyfully accept these new products, the so-called frankenfoods?

Well-being in many cultures, of course, is strongly based on beliefs of food. Food is central to the practice of the Hindu tradition; next to a wedding, it is this topic that commands the most space and energy of the writers of the texts on dharma (righteousness, duty) composed around the first millennium CE.

When we consider Hindu attitudes toward food in general and GM food in particular, we have to keep in mind that some attitudes may arise from specific Hindu practices. Other attitudes may emerge from Hindus' involvement and participation in global narratives, structures, and movements and their ethical take on any number of issues such as animal rights, multinational corporations, GATT, and so on.

One has to also explicitly keep in mind the diversity of Hindu communities and traditions; Hinduism is pluralistic with thousands of communities, castes, philosophies, and religious leaders. Caste, community, economic class, and the geographic area from which one hails have specific bearing on one's food tastes and habits. Brahmins from some areas are strict vegetarians, but some Brahmin communities from Mangalore or Bengal may eat fish. Some communities may belong to one of the mercantile castes, but if they are followers of the deity Vishnu (*vaishnava*), they may not only be strict vegetarians but completely eschew garlic and onions, among other vegetables. We will keep this plurality in mind when we talk about some Hindu attitudes to food.

In this chapter, I will consider the various ritual contexts in which food is used in the Hindu traditions, particularly in the classification of "pure" and "impure" foods. These ritual contexts and the classification are both significant because although Hindus may eat what some texts consider to be "impure" foods on an everyday basis, food taken on certain holy days, food offered to the deity, and food distributed as a mark of divine favor in temples should only be "pure" foods. These contexts, therefore, would determine not just if GM food is acceptable, but *when*, and in what circumstances, it may be acceptable.

There is no single central authoritarian structure within the Hindu tradition or even a handful of authoritative organizations that can give a considered opinion on whether a particular food is to be accepted or banned. Nor is there any single text that can be the sole seat of authority. As with many other issues in the Hindu traditions, there are a number of places where one can go for guidance, including but not limited to, community customs, texts, teachers, authority figures, and one's own conscience.

Centrality of Food

The centrality of food both in the spiritual texts and in the practices of many Hindu communities is most striking. Early Vedic sacrifices, more than three thousand years ago, involved offerings of food to various *devas* or deities. In the Upanishads—the philosophical sections of the Vedas that were composed around the sixth century BCE—there is a famous chant that is still recited every week in Vaishnava temple liturgies in many parts of the world. In this, the priest or the worshiper says: "This one who is in the person, and that one who is in the sun, are one: and the one who leaves this world knowing this goes up to the self made of food, goes up to the self made of mind. . . . He moves about the worlds, with food at his desire. . . . He continually sings this *saman*: 'Oh bliss . . . ! Oh, Bliss! Oh, bliss! I am food, I am food, I am food, I am the eater of food . . . I am the maker of the verse. . . . I eat . . . food and the one who eats food, I have overcome the whole universe. I am light like the sun' "[1]

This complex verse has many interpretations; we only note it here to show how central the idea of food is in philosophical and liturgical life. In later devotional poetry, sometimes the deity and the devotee speak of consuming each other, swallowing each other, and holding each other in their stomachs. Food, then, is not just central in everyday life but also a leitmotif in literature where the poets speak of having a religious experience.[2]

Hindus also offer food to the deities in temples and at home altars every day, sometimes several times a day. This food is consumed by devotees as a token of divine grace, and the blessings are literally ingested.

Despite this significance in many spheres, food is also important by its absence. Many Hindus fast on a regular basis. There are weekly, fortnightly, monthly, and annual fasts of various kinds and textures ranging from an absence of a particular product—meat, rice, grain, or any other substance—in the meal to a complete fast of not eating or drinking anything for a certain number of hours or even a whole day. These periods of fasting and feasting are connected to the lunar calendar, which is adjusted to the solar calendar. It is important to keep these ritual patterns in mind, because what may be "allowable" food on ordinary days may be completely avoided during the ritually significant days. These issues are significant in forming attitudes toward GM food. While some texts give details of the feasts and fasts, almost all the details are derived from local customs.

To understand the attitudes toward GM foods, we must also understand the many Hindu attitudes toward family and genetic inheritance. Biological descent and connection are central in both text and practice. People are identified as being from a particular extended family, a descendent of someone, or as part of a caste or social community. Patrilineal descent is particularly significant. In Tamil verse, poets speak of God saving seven generations of a family.[3] This attitude toward biological connections would be directly related to GM food. One participant in the focus groups conducted as part of this study said, "[O]ne gene, two genes . . . it does not matter," adding that GM food is "not pure food." Another participant said that "even one gene" would make it suspect; and yet another said that "the presence of one foreign gene makes it no longer satvic."[4]

The nature of the food, who touches it, and who cooks it were all significant in orthoprax families; today, they are of concern if they are used as ritual offerings or during worship. Traits are passed on not just biologically; there are other forms of biocultural mapping. Thus, according to oral tradition in many communities, the characteristics of the cook are supposed to pass on into the food. One may not find this notion necessarily in texts, but these concepts are certainly present in oral tradition.

Texts, Custom Narrative, and Practice

Perhaps one of the most unusual aspects of Hinduism—a tradition that has thousands of religious texts in Sanskrit and dozens of vernacular languages—is that custom and practice are equally, if not more, important than the sacred texts in many communities. Indeed, most Hindus are more familiar with narratives and rituals than actual texts. This was reiterated in the focus group where an individual said, "I haven't found any documented statement with respect to that myself, or maybe read it fully."[5]

In the past, legal rulings were made based extensively on local practice and contexts.[6] This aspect was missed by many of the early Western scholars of the tradition who described Hinduism from the content of their books. For example, the social divisions of society known as the caste system have been portrayed as being fixed and rigid, and other studies have depicted women as being totally dependent on men; but neither quite reflected social reality. Hindu texts, therefore, have to be balanced with customs and practices. These are considered to be so important that they are recognized by the Supreme Court of India to be as significant as the Sanskrit and vernacular texts that speak about ideal, normative behavior.

Thus, though there are many rules about food consumption, the only ones that matter are those that one learns from one's family, a neighbor, or a friend. In the past it was important to follow rules from one's own community, and these were significant in identity formation; in recent times, one learns from a number of sources, including the internet, and one is a "Hindu" in different ways, with different sources of authority, than say, one's grandmother. Hindus encounter and embody these narratives and practices through their relationships, both synchronic and diachronic, with families, communities, nature, and deities. These traditions are all part of what I call "custom narrative and practices" and are deeply connected with the embodied cosmologies found in Hinduism. Every Hindu community, family, and individual actively chooses and adapts stories and practices, *customizing* them, making them relevant and appropriate to the place and time. Most people will tell you that turmeric and neem have antiseptic properties and are powerful healing medicines, but today, some people may opt to buy them in ointment form in local pharmacies for external application or in a processed form in the local grocery store for cooking. The customary knowledge has been adapted for urban dwellers who prefer their traditional medicine from a local drug store; similarly, traditional knowledge is periodically customized to fit dwellers in Mumbai, New York, or Singapore.

The Hindu traditions are also known for multiple layers of interpretive narratives, many of them oral. The practice of not eating beef, for instance, may be explained in many ways; some may say that this form of meat is inherently bad for human beings; some may say that the killing of a cow involves violence; some may say that the cow gives milk and is therefore like a mother; some may say that Krishna, an incarnation of Lord Vishnu, was a cowherd and protected cows, so it behooves us not to eat beef, and so on. No single reason is considered to be right or wrong in itself in these practices.

Thus the many rituals, practices, and narratives pressed into use for well-being and healing may have some origin in texts, but it is not through these texts Hindus primarily learn about their practices. Rather, in the case of food, one sees it at home, remembers it was done by one's family, and then continues a family tradition. One may acknowledge it is in the text or that a holy person authorized it, but the details of such authority are usually not very clear. Custom narrative and practice are both textual and oral.

These customs, among other reasons, make the loosely structured Hindu traditions fluid and adaptable. People accept new customs, practices, even foods, and co-opt them very fast. These then become accepted tradition, and within a few generations it is impossible to see how people lived without these foods or traditions. Two examples are potatoes and fiery red peppers, both of which were introduced into Indian cuisine by the Europeans. Today, one cannot imagine the generic curry without these ingredients. Hindus can adopt new traditions and new foods quickly, and one can be Hindu in different ways.

Textual Resources

The word *dharma* appears in the early Vedic texts several times. In many later contexts, it means "religious ordinances and rites," and in others, it refers to "fixed principles or rules of conduct." In conjunction with other words, 'dharma' also means "merit acquired by the performance of religious rites" and "the whole body of religious duties."[7] Eventually, the prominent meaning of 'dharma' came to mean "the privileges, duties and obligations of a man, his standard of conduct as a member of the Aryan [i.e., noble] community, as a member of one of the castes, as a person in a particular stage of life."[8] Texts on dharma both described and prescribed these duties and responsibilities and divided up the subject matter into various categories. And it is in these books of dharma that one can look for some early perceptions on food in the Hindu tradition as well as for rules that make possible the dynamism within the larger culture. It is this flexibility that enables choices and decision making about how one leads one's life, one's involvement with the environment, and one's consumption of certain foods.

Sources of Dharma

By the first centuries of the Common Era, many treatises on the nature of righteousness, moral duty, and law were written. These are called the "*dharma shastras*" and form the basis for later Hindu laws. The most famous of these is the *Manava Dharmashastra*, or the *Laws of Manu*, which were probably codified around the first century and which reflect the social norms of the time. The *Laws of Manu*, along with some other texts, list the foundations for our understanding of dharma. *Manu* 2:6 lists these as the Vedas (*sruti*); the epics, puranas, and other *smriti* literature; the behavior and practices of the good people (*sadachara*); and finally, the promptings of one's mind or conscience. Variations of this list are found in the earliest texts of dharma as well.[9] Although these texts deal with common topics, they vary in their opinions and provide plenty of room for interpretation. Since these rule books were written by brahmin men for other brahmin men (a very small but influential percentage of society), they were by no means followed widely. Norms differed all over India according to caste, area, region, gender, and age.

The lofty ideals of the texts of dharma are, however, made accessible in the stories of the epics and puranas. Hindus in India and the diaspora know these narratives are better known than the *dharma shastra* texts. They understand stories from these texts as exemplifying values of dharma and situations of dharmic dilemmas. Rama followed his filial path and went into exile; Sita is the paradigm of *stri* (womanly) dharma, and so on. The *Mahabharata* is filled with narratives of dharmic dilemmas that add to the dramatic unfolding of the story.

Dharma and Classification of Food: Purity and Pollution, Auspiciousness and Inauspiciousness

While the concept of dharma as sustaining the universe is important in many texts, its interpretation in everyday life is of more significance. Although people do *not* ordinarily use 'dharma' in daily life to speak about their actions, their code of behavior is probably governed by the general and particular rules of righteousness found in the texts and in local custom. Examples may include the cooking of the correct kinds of lentils on ritual occasions, contemplating issues of war and peace, thinking through the ethics of environmental problems, and a consideration of new reproductive technology.

Complicated rules of ritual purity and pollution and dietary rules are seen in the many Hindu communities in connection with diet and medical therapies. Hindus from different communities from various areas of the world divide food into many categories. While an overall taxonomy of food is not possible here, we may note that many Hindus divide food into pure and impure. The classification into pure and impure may depend on the caste, community, and region of the Hindus. Sometimes the pure/impure categories are connected in some parts of northern India with what are called "raw" (*kachcha*) foods and the "ripe" or "correct" (*pakka*) foods. Other Hindus may further elaborate the pure/impure foods into a tripartite category. There are other categories as well: certain kinds of foods are considered "auspicious" because they are cooked for happy, life-promoting events, and others are inauspicious or those cooked for death-related occasions. Still another way of slicing this pie is to relate food to local, customary notions of the body and well-being. In some areas, certain kinds of food are seen as heating or cooling to the body; other notions follow a tripartite medical division of what are popularly known as *vata*, *pitta*, and *kapha*.

Religious practices and attitudes toward food in many castes within the Hindu traditions are marked by notions of "purity" and "impurity." Certain foods are ordinarily considered to be pure; these include many, though not all, dairy products. Milk, buttermilk, and yogurt, for instance, are "pure," but cheese is not. In general, egg, fish, fowl, meat, and liquor are considered to be impure foods, and only some vegetables and spices come under the "pure" category. In addition to avoiding meat and fowl, garlic, mushrooms, and several other items are not considered to be part of a pure diet by many followers of Vishnu; early texts on dharma have a blanket decree against these foods. There are also other criteria for purity and pollution; in many cases, purity is defined by the freshness of the food and when it is cooked. Until refrigerators became popular in the 1950s and '60s, leftover food was considered not fit for consumption. This is an example of how certain kinds of food considered as unacceptable in many communities came to be accepted. This acceptance came about in part because of advances in technology and the availability of consumer appliances such as the refrigerator and freezer. Food tasted by someone or food eaten from a plate becomes "impure," and another person cannot eat from the same plate. Most Hindus even today cannot stand the idea of someone lifting up the ladle, tasting the soup or sauce, and putting that same ladle (now impure because of contact with the taster's mouth) into the cooking pot. And it takes a while for many Hindus to accept a common dipping bowl for chips at a party.

In some parts of northern India, food that is cooked in or with water would be considered to be inferior (*kachcha* or "raw") to that cooked in fat, particularly in *ghee* or clarified butter. The latter was considered to be *pakka*, that is, complete or perfect. The translations of the concepts as "raw" and "perfect" are misleading because they do not refer to "cooking" in the ordinarily accepted sense of the word but rather to a range of specific treatments to which the food is subjected. Members of the communities that hold these concepts used to offer only *pakka*, or fried foods, because they were considered to be more pure to people of the so called "higher" castes. In another classification, well known among the many Hindu communities, a tripartite division of qualities, spoken of in many early philosophical traditions such as the *Sankhya* and texts such as the *Bhagavad Gita*, is used to understand food. Food is understood to have qualities that would give rise to similar attributes in the consumer.

In popular understanding, and by fitting a template from philosophical discourses on what we eat, food, like people and even deities, is said to have propensities to three characteristics: purity (*sattva*); sloth and stupor (*tamas*); and energy and passion (*rajas*). While this is not clearly discussed in the early texts on dharma, many Hindus try to fit in food regulations with these categories. Thus by eating certain kinds of foods, purity (*sattva*), passion and energy (*rajas*), or sloth and stupor may be highlighted in a person. We find these concepts voiced by many in the focus groups; the people in these groups said what one may hear in many Hindu families: one person said, "You become what you eat," and another, conflating various ideas, said, "The *Gita* says there are four types of people, and certain people become what they eat. Like if you are a meat eater you become very uncaring of others, and you're only interested in eating, and your mind is always on eating." A powerful statement came from a member of the focus group who said, "If I eat things I don't want to be, then my soul will be at risk."[10] Perhaps more than any other form of classification, this one may be pertinent to GM foods. Ordinarily, onions, garlic, fowl, fish, and meat would come under the categories of *rajas* and *tamas*. If genes or a genetic sequence from one of these was to be introduced into "pure" foods, it would almost certainly be considered to lose its purity.

In general, one may say that if a ritual is closely connected with Vedic mantras and is of some antiquity, the purer the food. Some Hindus understand this from the viewpoint of "authenticity." Some vegetables such as cauliflower, or potatoes, and some spices, such red pepper, are perceived to have been introduced by colonial powers. Potatoes and chilies have been introduced in India only in the last few centuries, yet it is impossible to imagine Indian cuisine without these items. Many vegetables, such as cauliflower and peas, are still called "English vegetables" in south India because the British are perceived to have introduced them. Thus, while all these are regular staple food in the everyday diet—it is, in fact, almost impossible to imagine Indian curries without these vegetables or spices—many of the "higher" caste Hindus may avoid them during ancestral rites (*sraddha*) or death rituals. Some rituals, such as ancestral ceremonies and the investiture of a sacred thread for a young boy, necessitate the cooking of the purest food. In this context, it means a throwback to earlier times when certain vegetables and spices were not around. This is particularly true for ancestral rites when great care is taken to ensure that jaggery/molasses and black pepper are used rather than, say, sugar or red chili peppers.

Another axis of classification would be that of auspicious and inauspicious foods. Some foods, by tradition, are cooked only on happy occasions and others, for rituals connected with death. Thus, some communities may use certain kinds of lentils, vegetables, and spices for weddings and others for death rituals. Purity and pollution do not overlap completely with auspiciousness and inauspiciousness; all lentils are considered to be *pure*, but only some varieties are used for auspicious events such as weddings and others for inauspicious events.

This segmented understanding of food in Hindu cultures is important for us; if GM food is accepted, it would be only for certain occasions and would be abjured for consumption on other ritual days. On some ritual feast days, some Hindus may feast with chicken, but on other ritual days (such as days when one is performing ancestral rites), one may only have the "pure" foods. Similarly, food offered to the deity, in most (though not all) cases, is "pure." Some "village" deities consume meat and quaff liquor, but this is not considered normative from the viewpoint of other communities.

Some of these classifications of food are regionally known or specific to communities; others follow categories better known through an Ayurvedic framework. There are sometimes fuzzy boundaries between foods consumed in regular diet and those taken for health reasons. Loosely following various forms of health-related traditions, Hindus may divide food into those that have propensities to restore balance to the various agents in one's body. There is no single theory governing these issues of balance. In some regional classifications, foods may be thought of as "heating" or "cooling." In others, the combination of what is known as the "five elements" in many schools of Indian thought is perceived to activate certain tendencies in one's body. The five elements are air, water, earth, fire, and what is commonly called "ether" or "space." Combinations of these give rise to tendencies called *vata*, *kapha*, and *pitta*. These terms cannot be translated easily. Borrowing upon medieval European thinking, colonial scholars called these three tendencies bodily "humors." *Vata* is loosely connected with air and is connected with a number of bodily activities including breathing; *pitta* is connected with digestion, among other functions; and *kapha*, related to water, is said to be connected with the lungs, sinuses, and so on. Imbalances are common; the right combinations of foods are said to restore the balance. Thus, certain combinations increase or decrease *pitta*; colds and sinus disorders may be combated with foods that have more "heat" in them, and so on. The question that would rise in connection with GM foods is whether we know how they will react in combination with other foods. As we know from many systems of medicine, sometimes one only needs a tiny portion of a particular ingredient or chemical to change the complexity of food.

The greatest amount of time in several texts of dharma is spent on listing forbidden foods, which varied through the different time periods and between authors. Although there is a common perception that most Hindus are vegetarians, most actually consume fish, fowl, or meat. The normative language, however, privileges vegetarianism; in India people are considered to be either vegetarians or "nonvegetarians" even though people in the first category are much smaller than the latter. Almost no one is a vegan. Some scholars hold that most people in India ate meat, possibly even beef, possibly up to the beginning of the Common Era. It is a matter of some controversy whether Indians ate beef during the time of the Vedas and whether the cow was a protected animal; however, it seems to be fairly well accepted that most Indians ate other kinds of meat and fowl then. Pandurang Vaman Kane, who wrote a multivolume treatise on Hindu dharma, remarks that it is quite remarkable that a civilization in which perhaps much of society consumed fowl or fish began to consider nonviolence and vegetarianism as the norm by the first millennium CE.[11]

Almost all rules, however, can be eschewed for medical purposes. Although garlic and onions are not "pure" foods, their usage is strongly encouraged in medical manuals for various ailments. Other similarly classified foods not considered to be right for ritual occasions may still be used for medical reasons.

Pure and impure foods and auspicious and inauspicious foods are different. Hindus grow up with various degrees of dietary regulations depending on their castes, religious communities, and degrees of secularization, and the guidelines discussed above form part of one's lifestyle. With accelerated modernization and secularization, these dietary regulations have been relaxed considerably in many Hindu households. However, even though many food regulations are lost because of practicality, convenience, and migration, a fair amount of them linger on and are specially pressed into use on ritual occasions or when one is suffering from ill health.

Right eating is not just what one can eat or avoid; in the texts on dharma, as well as in orthoprax houses, it involves issues such as the caste and gender of the cook (preferably male and high caste, or the lady of the house, except at times when she is menstruating); the times one may eat (twice a day, not during twilight times, not during eclipses, and a wide variety of other instances); not eating food cooked the day before; and so on. In earlier times, other directives were also in vogue; in detail, some of these equal or even surpass those given in many Confucian texts. They include the order of food courses in a meal; the direction in which the diner must sit (preferably facing east or north); how much one may eat (the number of morsels depended on the stage of life); the materials with which the eating vessels should be made; and what is to be done with left over food. Many of these directives and others common to the local region were followed for centuries.

There were several strict rules on whom one may dine with (best to dine alone!). Silence was recommended for the time of dining except to enquire after a guest's needs. Most texts say, and this was followed until probably the midtwentieth century, that one may dine only with people of the same caste and with people one knew. It was believed in many circles that one shared the sins of the people one dined with, especially if one sat in a single row with them. Through the centuries, we may see Hindus from many communities visiting shrines of other religious traditions; but many of them seldom ate with anyone other than their own caste and community.

The distribution of food was and is still a very important act of dharma. Many temples, monasteries, and educational institutions have *annadanam* ("donation of food") schemes to which one is exhorted to contribute. The walls of many temples are inscribed with exemplary acts of generosity by men and women who endowed monies for the feeding of pilgrims and religious personnel. This food was initially offered to the deity, deemed to be blessed by that god or goddess and distributed as a token of divine blessing. This food was and is called "*prasada*."

Prasada

Prasada (literally "clarity," used here in the sense of "favor" of the deity) is a generic term used for various offerings to the deity. Apart from the institutional offerings we noted above, individual tokens of worship are also offered to the deity. After first being offered to the deity, the flower or fruit that a devotee brings to a temple is returned as "blessed" and as holding the favor of the Divine Being. Such a piece of blessed food is usually *prasada*. This act of devotion, of seeing the deity and offering something and getting it back in a blessed state, is the simplest and most popular among votive rituals.

In many large temples, the deity is woken up in a similar manner in the morning, frequently to the sounds of music and Sanskrit prayers. Large quantities of food are offered to the deity at regular intervals.

If it is a Vishnu shrine in the south, the worshiper may be given a spoonful of holy water to sip and some *tulasi* (basil) leaves sacred to Vishnu. The devotee then puts a pinch of the basil in his/her mouth to ingest. Occasionally, some fragrant sandalwood paste from the deity's body or flowers that adorned the form of the goddess may be given to the worshiper. The devotee receives this *prasada* with his or her right hand (the left hand is considered polluting) and ingests that divine grace.

It is in all these ritual contexts of offering food to the deity and redistributing it to devotees, in cooking on some occasions such as ancestral rites or the investiture of a sacred thread, that Hindus do not use any "new" or "impure" foods. So, even if some Hindus regularly ate fish or chicken, on these occasions, they would avoid GM foods. Using products that involve animal to plant transfer—that is, transfer of genes between species—would be unacceptable during these times.

Feasting and Fasting

The modes of celebrating festivals and domestic rituals change over the centuries and are largely based on custom narrative and practices. They reify sectarian, linguistic, regional, and transnational Hindu identities, transform, and are transformed by, social, economic, and political forces.

The focus of some festivals is on food. Pongal, a festival celebrated in mid-January in south India, for instance, is the name of the celebration as well as a particular kind of food.

The second day of this festival (usually January 14 or 15) is the actual day of the Pongal celebration. The cooking and consuming of the Pongal dishes are common to all Tamil households. The two kinds of *pongal*—sweet and peppery—are made. The sweet *pongal* is made by cooking a specific kind of lentil (*mung dal*) and rice in milk and clarified butter along with cardamom, raisins, and cashew nuts. The immense popularity of this dish can be seen from the dozens of recipes posted on the internet.

The *ven* or "white" *pongal*, which, in some south Indian houses is made every day of the Tamil month of Markali, which precedes the festival, is also made on Pongal. Rice and lentils are flavored with ginger root, cumin, and pepper corns and laced with clarified butter. A mildly tangy dish, it is said to contain the spices that reinforce one's sense of well-being on the cool dawns of the month of Markali. According to oral tradition, all the spices mentioned here are supposed to be "heating" in nature and help one fight phlegm, colds, and so on. What is significant about this explanation is that many Hindus do not just cook a particular dish on a certain festival day but tend to pay attention to the appropriate nature of that food to the weather and to one's physical condition during certain seasons.

Fasting

Most Hindus pray explicitly for the fulfillment of specific wishes and make various pacts with the deities. If the wish is fulfilled, or in order for it to be fulfilled, they may go on a pilgrimage, shave their heads, have dietary regulations (various kinds of fasting), or perform an assortment of rituals at home. Some of these acts are gendered: women do specific kinds of fasting for the well-being of the families or for a long life for their husbands. The votive rituals, which are more frequently undertaken by women and which involve fasting, provide a sense of empowerment for women and simultaneously make the celebrant a recipient of the deity's grace.[12] They are done to manipulate and give a positive spin to one's karma and achieve one's desires. They affirm the connections between family members and close friends, on the one hand, and deities and devotees, on the other hand, creating reciprocal networks of devotion and favors. The periodic fasting and feasting by women, with an attendant retelling of narratives and performance of rituals, is undertaken for the well-being of the family, good health, and other specific purposes.

There are many kinds of fasts undertaken both on regular days of the week, lunar month, or calendar year, and others that one may do at one's will. In a few—and these are not very common—the participant may fast completely for part of the day or even the entire day. In some, such as the festival of Navaratri, women from Maharashtra may avoid grains completely. Certain vegetables that are ordinarily consumed—such as bell peppers, eggplant—may be avoided on days of fasting. There are reasons for some of these customs but not for all. It is said that eggplant is avoided because it has a number of seeds, and eating seeds terminates potential life. One may eschew certain kinds of vegetables that are considered "foreign"—introduced in the last few centuries. Others may be on a liquid diet. On days when there are worship ceremonies to the goddess Santoshi Ma, a relatively new goddess in the Hindu pantheon, the entire family avoids anything that has even the slightest sour taste. There are as many fasting sequences as there are families.

What Trumps Traditional Notions of or Ritual Attitudes to Food?

What would be the conditions under which Hindus would accept modified food—whether it is GM or otherwise? We do know that Hindus have been open to "foreign" foods. The pace of acceptance was slow up to the twentieth century and accelerated from the latter part of the twentieth century. Food tastes have changed along with growing awareness of not just various parts of India, but of the world, and also of increasing consciousness of health benefits. Until even the 1960s, food habits were hard to change in India. South Indians were consummate rice eaters and would not eat *roti*, the unleavened wheat bread that was traditional fare in northern India. After the 1970s, cuisine from various parts of a state or from specific communities has taken on a life of its own and prospers alongside international cuisine.

Adventurous food thrives alongside ritually pure food and comfort foods. What makes a food acceptable? A quick look at these issues may help us understand how—if at all—GM foods are accepted.

The many rules of ritual pollution and purity and auspicious and inauspicious foods are trumped by a number of considerations: practicality and the feasibility of holding on to the rules, health considerations, an attention to "general" or "common" dharma, which includes virtues such as compassion and nonviolence, and devotion to the deity and other devotees.

Practical Issues and Vegetarian Diets

Practicality and availability as well as medical reasons may be important in the adoption of GM food. This includes the economic affordability of any item. If something is easy to procure, financially more affordable, *and* it fits into one's menu easily, it would be easy to adopt. Just as new foods were introduced and accepted, just as "toned" milk and skim milk are now part of the diet, GM foods too can be accepted. Hindus—and most Indians, for that matter—are only too keenly aware of population issues and have adopted various kinds of GM grains. But how about those foods that involve genes from another species—how about the salmon gene in the tomato? *Rasam*, the South Indian soup that contains tomatoes, would still taste the same, but the vegetarians would not want to use tomatoes. It is crossing the species—with animal genes in plants—that would bother vegetarians. But still, with easy availability and financial incentives, some vegetarians may accept GM foods for everyday consumption.

Health Consciousness

Medicine and well-being are important considerations in dietary matters. Dietary rules are followed in normal circumstances and as long as one enjoys good health; however, if one is ill, these norms are overruled in the preparation of certain medicines. Hindus have been careful about religious regulations but are very practical people as well. Charaka (c. third century CE), who wrote one of the earliest and most powerful texts on healing, puts good health and general well-being in the broadest possible context. He says that a human being should, above all, have the will to live. Next, the person should have a desire for prosperity, and finally, the desire for liberation from the cycle of life and death.[13] It is in this context that we can understand Hindu attitudes to food and health. One needs good health; ritual injunctions come only after this. Thus, although one is prohibited from eating garlic, when prescribed by a health practitioner as part of a medical concoction, a person may eat it willingly. If, therefore, there is proof that there are specific health benefits to certain kinds of food—those that have high protein, those that are fortified with essential vitamins, for example—GM foods would be acceptable. Energy drinks have become popular even when they contained egg products because the advertisements targeted parents who wanted their children to have the right foods. Thus, a member of the focus group asserts: "For purpose of curing any particular disease, like let's say AIDS research or cancer, for purpose of medicine, and for the purpose of scientific research for getting medicines for a disease like Crohn's disease, Parkinson's disease, and any such, we don't have medicines, it may be acceptable to use this therapy but not for daily consumption for normal human beings to eat."[14]

However, from all the literature—including the questions raised in many chapters in this volume—we realize that the claims of health benefits are questionable and that long-term fallout is not clear. A significant factor in acceptance of GM foods, whether one is Hindu or not, would be answers to these commonsense questions.

Devotion

In Hindu narratives, ritual rules about food are trumped by actions prompted by devotion, compassion, or generosity. These actions trump the rules of purity and pollution or other dicta of the texts of law and ethics *and* those that have been established by custom. Take for instance the story of Sabari in the *Ramayana*. A simple, unlettered devotee, Sabari, it is said, offered fruit to Lord Rama. According to Hindu text *and* custom, the food that one should offer the deity should not be previously tasted by a human being. As noted earlier in this chapter, once one eats from or tastes a dish, it is considered "left over" and said to be ritually impure. In social custom, the notion of "impurity" is strong: one cannot and should not taste food first and give from the same portion to a guest. Food that one tastes first is considered to be impure because it is said to be touched by one's saliva. And yet, the narrative goes, Sabari, in her eagerness to serve Rama only sweet fruit, apparently kept biting into each fruit to ascertain whether it was sweet or sour and then gave the half-bitten fruit to him. Although this act is considered to be ritually wrong and unacceptable according to the texts of law and practice, Sabari's devotion trumps the ideas of purity and pollution and is held as paradigmatic in the Hindu devotional universe.

The latitude offered by balancing texts with practice, and then both of them with acts of love or compassion, offers communities the space to celebrate actions that go contrary to the texts or to introduce innovations that may eventually take root. Based on our discussions in this chapter, we may raise the question: how then, will some Hindus prioritize the values by which they may accept or reject the use of GM food?

As we saw earlier, many Hindus would have a segmented acceptance of GM foods; they would consume them in some contexts and reject them in others. The hot button for many Hindus would be the introduction of what is regarded as animal genes into vegetables, fruits, or grains. If the origin, however remote, is an animal, then the modified food would be considered "impure." But it is not just animals; we cannot even put traits of onions or garlic into other food, because the modified food then becomes impure and unfit for ritual use. Similarly, genes from a grain into fruit would be troublesome to those Hindus who fast on some days and abstain from all grains.

The ritual context or the occasion in which the food is consumed will be important. Thus, if GM foods have "impure" genes in them, they would almost definitely be rejected for ritual worship. During ritual worship they would have to be offered to the deities in the temple and then distributed to the devotees as *prasada*, making GM foods unsuitable. They would also not be used for meals in rituals that are perceived to have "vedic" components, such as ancestral rituals and thread investiture.

Hindus may also accept it if the alternatives of not accepting would be worse in their minds. Most Hindus who live in many parts of the world and many urban Hindus in India would eat cheese, and many vegetarian Hindus think of this as almost a staple when dining with in Western-style restaurants. Thus, if one is faced with the hypothetical situation where the cheeses available were made with animal-derived rennet and those that were produced with other "articifical" substances, they would almost certainly opt for the latter because they may believe it would help them be better Hindus.

Most Hindus who do eat fish, fowl, and some forms of meat may accept GM foods if it can be shown, as far as it can be shown, that it is not hazardous to one's health. It is, however, the health factor that would be a wild card in considering the issue of acceptance. If it is considered beneficial to health, it may be accepted. But there are so many forms of health care systems in India connected with the Hindu tradition—Siddha medicine, Ayurveda, and so on—not to mention concepts of *vata*, *pitta*, and *kapha*, hot and cold foods, that people would worry about the interaction of these foods with others. The origin of the gene is as important as the traits it may have. So even if you were to have a synthetic gene that contains DNA like that of a pig gene, it may not be acceptable because that may be the gene (albeit in interaction) that causes the characteristic that makes it an onion or causes it to be a "hot" food or a "cold" food.

Samanya Dharma: The Dharma of All Human Beings

While food regulations are part of dharma, the texts and narratives of dharma also celebrate human virtues: compassion, non-violence, giving, gratitude, generosity. A famous line on dharma goes: "Lack of enmity to all beings in thought, word, and deed; compassion and giving—these are the signs of the eternal dharma." Many Hindus think of nonviolence as important even if they are not vegetarians.

What then, if we were to argue that to feed the hungry of the world, to keep the bread baskets, full one should grow crops that are resistant to disease and pests, that can grow with very little resources (increasingly important in a water-deprived world), and that are so fortified with minerals and vitamins that no child would die of malnutrition again? Would it not be incumbent on all Hindus who proclaim the *sanatana dharma* to accept certain kinds of GM foods?

The answer would be in the affirmative if one can (a) demonstrate the long-term benefits and absence of health hazards and (b) avoid the introduction of genes from animals. For many Hindus even the second condition may be acceptable if the first one were met; they would simply avoid using modified foods with animal genes in religious contexts.

It is in this context that it is important to note that Hindu narratives are *not* against the notion of "genetic modification" per se. There are many stories about supernatural or "unnatural" means of conception and giving birth in the Hindu epics and *Puranas*. In the *Mahabharata*, one hundred embryos are grown in separate containers by a queen, Gandhari. In other texts, an embryo is transplanted from one woman to another; Krishna's brother Balarama is transplanted into another womb when still in an embryonic stage. In some narratives, women consume divine potions, children are born miraculously, and deities are invoked to "fertilize" the woman if the husband cannot procreate. Even though these tales that legitimate the new reproductive technologies are generally not invoked, the technologies seem to have been accepted easily for human reproduction. One can argue that the awareness of many stories such as these made it easier for people to adopt new reproductive technologies. It is also important to keep in mind that Hindus are not in general averse to science and technology; in fact, early Hindu culture showed great advances in many scientific fields. While human beings are important, the world was not anthropocentric. Even today, in many forums, the tradition co-opts science. Advances in astronomical knowledge and technological considerations do not make Hindus become skeptical of astrology; rather, they use the latest computers and write software programs for horoscopes.

There is, therefore, no knee-jerk rejection of scientific and technological advances, but frequently a segmented acceptance to come up with innovative responses. Nor do people worry too much about playing God or Goddess; everyone who has had an Ayurveda medication, allopathic antibiotic, or had an aspirin has in some fashion interfered with nature, interfered with karma. Every time we build a dam we play with nature.

Food, however, touches the core of lives. Even though plants have been bred for centuries, the pace and the extent with which such modifications take place are worrying. What if these do not deliver on the promise of "well-being" and instead accelerate health and environmental disasters? The Hindu acceptance or rejection of GM foods, therefore, would be based more on scientific reasoning, on the one hand, and persuasive arguments based on notions of dharma, on the other. These arguments could arise from a number of locales. Local customs and narratives would be useful—as long as such genetic modifications were within plants without introduction of animal genes—insofar as it is acceptable to talk about "animal" genes versus simple gene sequences. Acceptance or rejection could be based on the context of use: almost no Hindu would use GM food for religious ritual contexts, to offer to the deity, or to serve as prasada if they are known to contain genes from animals or any one of the forbidden fruits or vegetables. But for everyday use, the question of religious integrity would depend on ethical integrity. On the one hand, there is the promise of solving the problem of hunger and malnutrition. On the other, there are worries about long-term hazards and accusations of this feeding into giant multinational corporations swallowing small farmers and the poorer nations and modifying that capital into profitable portfolios. The traditional dictates of *samanya dharma* about nonviolence, compassion, and so on, can be pressed into an argument for GM food if only the issues of health and economic hazards can be addressed.

This is where the question of personal integrity, that which allows for the happiness of the self—what we loosely call "conscience"—comes in. This is where the fourth source of dharma, which we noted in the beginning of this chapter, can be pressed into use. It is this area that has allowed for the resilience and dynamism of innovative traditions within the Hindu traditions. Combined with avenues that offer health and fullness to the needy, it becomes a dharmic imperative to explore the promise of GM food. And even if it is not pure and fit for the deity, *if and only if*, these products are developed and distributed with economic integrity, compassion, and without health hazards, it is certainly fit for the devotees. The story of Sabari certainly assures us that when these acts are done with love, the rules can be bent. In a world with shrinking resources and an increasing population, that may be the only way in which all may participate in the traditional Hindu blessing: "May we live a hundred autumns; may we flourish for a hundred autumns."

Notes

1. *Taittiriya Upanishad III.10. 5* in Patrick Olivelle, tr. *The Upanishads* (New York: Oxford University Press, 1998), p. 114.

2. John Carman and Vasudha Narayanan, *The Tamil Veda: Pillan's Interpretation of the Tiruvaymoli* (Chicago: University of Chicago Press, 1989), pp.159–79.

3. In the weddings or ancestral rituals of some communities, for instance, seven generations of ancestors are mentioned. An example of seven generations of family members is seen widely in Tamil literature; see for instance, the ninth-century poem by Nammalvar, *Tiruvaymoli* 2.7.1.

4. Hindu Focus Group, February 6, 2005, Victoria, BC.

5. Ibid.

6. Richard W. Lariviere, "Justices and Panditas: Some Ironies in Contemporary Readings of the Hindu Legal Past," *Journal of Asian Studies* 48, no 4 (1989), pp. 757–69.

7. Panduranga Vaman Kane, *History of Dharmasâstra (ancient and mediaeval religious and civil law)* (Poona: Bhandarkar Oriental Research Institute, Volume 1, 1968), pp. 1–2.

8. Ibid., p. 3.

9. Patrick Olivelle, trans., *Dharmasutras* (New York: Oxford University Press, 1999). See particularly, Baudhayana Dharma Sutra 1.1.1–6 and Gautama's Dharma Sutra 1.1.1–6.

10. Hindu Focus Group, February 6, 2005, Victoria, BC.

11. Kane, *History of Dharmasâstra*, pp. 757–99.

12. See, for instance, Anne Pearson, *Because It Gives Me Peace of Mind: Ritual Fasts in the Religious Lives of Hindu Women* (Albany: State University of New York Press, 1996);Tracy Pintchman, "When Vows Fail to Deliver What They Promise: The Case of Rajavanti," Paper presented at the American Academy of Religion, Orlando, Florida, 1998.

13. Dominik Wujastyk, *The Roots of Ayurveda* (New Delhi: Penguin Books, 2001), pp. 61–62.

14. Hindu Focus Group, February 6, 2005, Victoria, BC.

8

The Karma of
Genetically Modified Food

A Buddhist Perspective

David R. Loy

What can Buddhism teach us about genetically modified food? Shakyamuni, the historical Buddha, lived at least twenty-four hundred years ago in Iron Age India. Needless to say, he and his followers did not know anything about the genetic structure of life, much less the possibilities of modifying it technologically. It is not surprising, then, that I have not been able to find references to genetically modified organisms in any Buddhist text—though I admit that my search has not been very thorough. The alternative is to extrapolate from traditional Buddhist teachings, especially (but not only) those pertaining to food practices, to see whether they might give us some insight into our present situation. These may help us determine what a Buddhist should or should not eat, but we also need to consider some of the larger issues that GM food raises—particularly the ways that new food technologies are being promoted.

At the beginning, however, we already encounter a problem: there is really no such thing as Buddhism. We need to take account of different Buddhisms. Buddhism is a missionary religion, but its message spread in a unique way: rather than challenge local deities and religious practices, Buddhism usually preferred to coexist, even merge, with them. For example, Chan (Zen) Buddhism is arguably as much Taoist as Buddhist; and Tibetan Buddhism developed out of the interaction between native Bon animism and Indian tantric Buddhism. The result is a variety of very different cultural traditions, so diverse that it is sometimes difficult to see what they share except for the label "Buddhist."

In accordance with this diversity, Buddhism has tended to adapt to local dietary customs rather than import and impose food restrictions. Given the difficult climate of Tibet, for example, it is not surprising that Tibetan Buddhists have usually eaten more meat than vegetables. Another factor encouraging this variety is that, in general, Buddhism has been less concerned about what we eat than how we eat it, since the *dukkha* (suffering or "dis-ease") that it addresses is rooted in our craving, and food is the second most popular example of human craving.

Nevertheless, there are some important distinctions within Buddhism and among Buddhists that have had important implications for food practices, especially those between monastics and laypeople, and between Theravada (South Asian) and Mahayana (mostly East Asian) Buddhism. By no coincidence, the distinction between Buddhist monks and the laity was raised at the very beginning of the Theravada Buddhist focus group discussion, for there are significant lifestyle differences. Monastics, in principle, focused on following the path of the Buddha and are expected to live a simple life largely unconcerned about mundane matters such as food.[1] In most Buddhist societies they eat only before noon (and usually only once). According to the *Patimokkha* that regulates their daily lives, "There are many fine foods such as these: ghee, butter, oil, honey, molasses, fish, meat, milk, and curds. If any *bhikkhu* who is not sick should ask for them and consume them, it is an offense entailing expiation."[2] Notice the careful wording. Evidently the problem is not with these foods themselves, but that desiring them and indulging in them is a distraction from what monastics should be concentrating on. There is no suggestion that lay followers should also avoid them, and the careful qualification—"any *bhikkhu* who is not sick"—exemplifies the pragmatic Buddhist approach: there may be times when even monastics would benefit from consuming them.

However, the main food issue for Buddhists has been, and continues to be, whether one should be vegetarian—somewhat complicated by the (contested) fact that, according to the earliest accounts we have, the Mahaparinibbana Sutra, Shakyamuni Buddha died of a stomach ailment caused or aggravated by eating pork.[3] Buddhist vegetarians have sometimes considered this fact scandalous and denied it, but it is consistent with what we know about the early Buddhist community.

According to the Vinaya, rules established and followed by the Buddha himself, Theravada monastics are mendicants. They do not grow or raise their own food; they beg for it. Being dependent on what is donated to them each morning, they are not required to be vegetarian, with two important restrictions. First, ten kinds of meat are prohibited, usually translated as bear, lion, hyena, tiger, panther, elephant, horse, serpent, dog, and human flesh—but not including beef, pork, or fowl, the types of meat most commonly eaten.[4] In each case, significantly, the prohibition seems to involve concern for the social or physical consequences for the monks themselves, not for the animals eaten.[5]

A more important restriction is often followed by Buddhist laypeople as well as *bhikkhu*: not to eat meat (or fish) if you know or have reason to suspect that it was killed for you. "If a bhikkhu sees, hears or suspects that it has been killed for his sake, he may not eat it" (Mahavagga, Vinaya Pitaka).

Why not? It seems a compassionate policy, given Buddhist emphasis on not harming living beings. However, the issue of animal suffering is cited in Buddhist texts less often than one's own karma: it is bad karma to cause the death of any sentient being. Even when those texts mention the importance of compassion, the main concern is often the negative effects of meat, ending on one's own capacity to cultivate compassion.[6]

I once heard a Buddhist teacher say that it is okay to eat meat, provided that it has passed through three pairs of hands before it gets to you. Presumably that is because the karma has worn off by then. This seems a rather self-centered attitude, taking advantage of the unfortunate situation of others who willingly or unwillingly have the job of butchering and processing meat for the rest of us to consume. Today the mechanics of the meat industry assure us that many hands have had a role in preparing our meat, but I think that does not necessarily resolve the important issue, from a Buddhist perspective. One might conclude that none of those plastic-wrapped chickens in the supermarket has been slaughtered for me, yet one can just as well argue, given the way the food industry functions, that any I might purchase has been slaughtered for me, because all of them have been raised and killed for all of us consumers.

Today there is a movement among expatriate Tibetan Buddhists (most of whom now live in the more tropical climate of India) to become more vegetarian, led by the Dalai Lama (who nevertheless sometimes eats meat for health reasons). This development is more consistent with a general Mahayana emphasis on vegetarianism, a concern especially strong in China and textually supported by well-known Mahayana scriptures such as the Lankavatara Sutra, the Surangama Sutra, and the Brahma's Net Sutra. These sutras, developed much later than the Theravadin sutras of South Asia, and East Asian Buddhism, were largely ignorant of the earlier teachings. The main points the Mahayana texts make in favor of vegetarianism are that eating meat

- was prohibited by the Buddha (according to the Lankavatara)
- is inconsistent with the first Buddhist precept, which prohibits taking the life of any sentient being
- produces bad breath and foul smells that inspire fear in other beings

• inhibits compassion and causes suffering to animals
• prevents progress in Buddhist practice and causes bad karma
  (e.g., you may be reborn as a lower animal), and . . .
• you may be eating a former relative.[7]

In accord with this, in the sixth century, Chinese Buddhism (unlike South Asian Theravadin Buddhism) began to emphasize vegetarianism. Chinese and Korean monastics today continue to abstain from meat and fish (often milk products and fertilized eggs, too), and, as the Chinese focus group mentioned, many devout Chinese laypeople are also vegetarian.[8] Curiously, it seems to have been the laity that played a leading role in this transformation: under the influence of Mahayana texts such as the ones mentioned earlier, as well as popular stories about karmic retribution, laypeople came to expect monastics to uphold higher standards of purity and renunciation. By the tenth century vegetarianism had become a minimum standard to be followed by all monks and nuns in China. As in South Asia, monastics are dependent upon lay support, so the concerns of an increasing number of lay vegetarians could not be ignored.[9]

The only other important dietary prohibition in Mahayana (also mentioned by the Chinese focus group) is to avoid the five "pungent odors," usually translated as garlic, onions, scallions, shallots, and leeks (sometimes chillies and other spices are added to this list).[10] In addition to the often-objectionable smells associated with them—perhaps the main concern in a crowded monastic situation?[11]—the Surangama Sutra claims that they are stimulants to anger if eaten raw and stimulate sexual desire when cooked.

Two points should be kept in mind regarding the above dietary restrictions. First, although monastics in principle have no choice, laypeople choosing to follow them make a personal decision in the sense that such practices are not required in order to be a Buddhist or follow the Buddhist path. Not observing them may create bad karma and make one's spiritual path more difficult to follow, but that is one's own decision. Second, as mentioned earlier, the key to Buddhist self-cultivation is less the "outer practice" of what one does than the "inner path" of how one does it. This is especially emphasized in Mahayana, which has a more relaxed attitude toward all such observances. *Upaya* (skillful means) may sometimes prompt us to break precepts in some situations, yet that may be okay, because Buddhist rules, like other teachings, are pragmatic rather than absolute. Since they have not been imposed upon us by an absolute deity, the issue is not sin or disobedience, but our *dukkha* suffering and the best way to alleviate that. A moral mistake is not an offense against God, but an unskillful act that causes more trouble for ourselves. Precepts are vows I make to myself, that I will try to live in a certain way, with the understanding that when I fall short then I will bear the karmic consequences. In sum, this generally involves a flexible attitude toward food practices.

Genetically Modified Food?

DHARMA REALM BUDDHIST ASSOCIATION
Headquarters: City of Ten Thousand Buddhas

Talmage, CA 95481–0217 USA

December 21, 1996

FOR IMMEDIATE RELEASE

BUDDHISTS CONDEMN GENETICALLY ENGINEERED FOODS

ADVOCATE LABELING

A major international Buddhist organization has formally condemned genetic engineering of food and advocated its required labeling. This is the text of Dharma Realm Buddhist Association's resolution:

Genetic engineering of food is not in accord with the teachings of Buddhism. Genetic engineering of food is unwarranted tampering with the natural patterns of our world at the most basic and dangerous level.

Lack of labeling of genetically engineered food is a de facto violation of religious freedom. Without labeling, Buddhists have no way to avoid purchasing foods that violate their basic religious beliefs and principles; and Buddhist vegetarians have no way to avoid purchasing foods that contain genes from non-vegetarian sources.

All countries are urged to require labeling of all genetically engineered food.[12]

What does Buddhism (or rather, what do Buddhisms) imply about genetically modified food? According to the Dharma Realm press release (above), the answer is clear: "Genetic engineering of food is unwarranted tampering with the natural patterns of our world at the most basic and dangerous level."

One can share this concern about tampering with nature—as I do— but there is nevertheless a problem in claiming that GM food does not accord with Buddhist teachings: there is virtually no support for the view that "unnatural is bad" in any important Buddhist text, because Buddhism has not valorized nature or "being natural" in the way that the West has often done.[13] The notion that "it's best to be natural"—a general principle important in the lifestyle of many Western converts—may or may not be wise, but it is not Buddhist. Contrast the Dharma Realm press release with the following report on Asian Buddhist attitudes toward stem cell research taken from an article by Jens Schlieter titled "No Buddhist Hard Line on Stem Cells" in the April 2004 issue of *Science & Theology News*:

> While many Buddhists in the Western countries view any destruction of embryonic life as wrong regardless of the possi- bility that a cure for severe diseases like Alzheimer's or critical spinal cord injuries might be found in a distant future, a con- siderable amount of Asian Buddhists come to a more permissive conclusion. First of all, *Buddhists are not inclined to see a man- made creation as something competing with a "good" nature*. There is a very positive attitude toward changing nature's course if it enhances the welfare of all living beings, and more so if it allows medical advancements. . . .
>
> Second, *Buddhist ethics are not principles to be followed as law*. They are not designed as expressions of indisputable human rights or as a consequence of dignity inherent in every human being. Ethics are much more a matter of personal choice; prin- ciples like the one of "non-harming" should be followed as guide- lines, and in extraordinary circumstances need not be applied in the strictest sense.[14]

Although food is not the issue addressed in Schlieter's article, its impli- cations apply to GM food as well. Despite the Dharma Realm press release, Buddhism generally has had a nonnormative understanding of nature, which does not appeal to "natural law" or some similar standard that must be observed.[15] As Lambert Schmithausen has concluded from his examination of ecological ethics in the early Buddhist tradition, the early texts empha- size the beauty of nature less than the struggle for life, the prevalence of greed and suffering, and most of all the universality of impermanence and decay.[16] Our distinctively Western ambivalence between infatuation with technological progress and romanticist nostalgia for a return-to-nature is un-Buddhist, because Buddhism does not assume a duality between them. Asian Buddhists are unlikely, therefore, to object to GM food on those grounds.

The Dharma Realm press release goes on to claim that lack of label-ing "is a de facto violation of religious freedom" because it becomes very difficult to avoid GM foods, a particular problem for Buddhist vegetarians who may want to avoid nonvegetarian genes in their vegetables. In contrast, members of the Theravada Buddhist focus group, although agreeing that it is important to know and have a choice, also agreed among themselves that there is nothing in the five precepts[17] that suggests a scientist should not take a gene from one species and transfer it to another one. They said that genetic modification "is not a big problem . . . if it is going to improve the food" and if "it's for the whole world—the human race." Most of them would eat a tomato with an "antifreeze" fish gene: there is "nothing in the Buddhist perspective that says the natural tomato is the right tomato." As they pointed out, focusing on "naturalness" would be inconsistent with Bud-dhist emphasis on the impermanence of all things.

Nevertheless, an exception needs to be noticed. Some members of the Chinese focus group did mention that Taoists "believe in nature," and "any-thing grown should be natural without interference," and a concern similar to that of the Dharma Realm Buddhist Association (a Chinese institution) was also expressed when one of the Chinese group members mentioned a Buddhist friend who "has great difficulties with eating a tomato that has a fish gene in it because that violates what she is practicing. And she feels it's really wrong for that [option] not to be presented to her." As for many of the people in the other focus groups, the main issue for her seems to be animal-to-plant gene transfer, especially problematic for vegetarians.

One can accept that lack of labeling violates her right to have a choice, regardless of her reasons for demanding it, but from a Buddhist perspective a question still remains: why does a single fish gene violate her Buddhist vegetarianism? The usual Buddhist objections to consuming meat and fish—cruelty to animals, bad karma, violating the first precept, bad breath, eating a relative—can perhaps be avoided, depending upon how the genetic modification is conducted, yet it is not clear whether that would satisfy the (absent) Buddhist friend or why it might not. Part of the difficulty may be due to ignorance of genetics: implanting a fish gene sounds like putting a piece of fish in the tomato. Reading the transcript of the Chinese group, however, I suspect that the objection is based more on feelings than reasons. Is the problem that such a tomato is unnatu-ral? For the Burmese Theravada Buddhists, a genetically modified chicken for example—so enlarged for breast meat that it could not walk[18]—clearly would be unacceptable, not because such a chicken is unnatural, but because such a chicken would suffer. For the Chinese group, including its Buddhists, transferring genes between plants was probably okay as long as onion genes and so on were not used. "But where do you stop?" one person asked. "It's impossible to draw the line."

The simplest explanation for the difference between these concerns brings us back to the point about cultural interaction emphasized at the beginning of this chapter: the Burmese Theravadins are more consistent with the earliest teachings, which do not privilege "the natural." In contrast, Chinese Buddhism has naturally (!) been somewhat influenced by traditional cultural values that emphasize harmony within society (Confucianism) and with nature (Taoism). This, of course, does not make Chinese Buddhism any less authentic as a type of Buddhism, yet it is important not to conflate it with Theravada Buddhism. Does this mean that from a Chinese Buddhist perspective genetically modified food should not be consumed, because it is unnatural? Again, this point was not made explicitly during the Chinese focus group discussions, but neither was there clear denial of such a concern. We are reminded that there is no Buddhism, only Buddhisms. Where do those differences leave us?

The *Cetana* of GM Food

Both the Theravada and Chinese focus groups expressed concern about the motivations behind the introduction of genetic modifications into food. According to a Chinese participant, "It's not just about scientific capability but whether we should do it." In explaining why we should not eat animals that have been slaughtered for us, Theravada participants highlighted the importance of *cetana,* or "intentional action," in Buddhist teachings. This emphasis on motivation and intention points at what is distinctive about the Buddhist perspective, so in the rest of this chapter I shall try to clarify that perspective and what it implies for genetically modified food. This involves expanding the evaluation criteria beyond the narrow issue of traditional dietary restrictions—as we have seen, not such a major issue for Buddhism—by considering broader issues about how consistent GM food is with Buddhism's basic worldview and understanding of human motivation. What role is the introduction of GM food likely to play, if any, in our individual and collective struggles with *dukkha*? To evaluate the ethical implications of genetic modification, we first need to understand the Buddhist understanding of karma, which was revolutionary in the way it emphasized *cetana.*

Earlier Indian teachings such as the Vedas usually understood karma more mechanically and ritualistically. To perform a sacrifice in the proper fashion would lead to the desired consequences. If those consequences were not forthcoming, then either there had been an error in procedure, or the causal effects were delayed, perhaps until one's next lifetime (hence the need for reincarnation). The Buddha transformed this ritualistic approach to controlling one's life into an ethical principle by focusing on our motivations. "It is *cetana*, monks, that I declare to be karma. Having willed, one performs an action by body, speech and mind."[19] What distinguishes our actions from mere behavior is that they are intended.

To help us understand how this innovation *ethicized karma*, and how it is relevant to our food practices, it is helpful to distinguish a morally relevant act into its three aspects: the *results* that I seek, *the moral rule or regulation* I am following (or not following), and my mental attitude or motivation when I do something. These aspects cannot be separated from each other, of course, but we can and often do emphasize one more than the others. In modern moral theory, utilitarian theories focus on consequences, deontological theories focus on moral guidelines such as the Golden Rule, and "virtue theories" focus on one's character and motivations. This may seem abstract, but it is quite helpful. In the Buddha's time, the mainstream Brahmanical understanding of karma emphasized the importance of following the detailed procedures (rules) regulating each ritual; naturally, however, the people who paid for the rituals were more interested in the outcome (results). When we evaluate genetically modified food, emphasizing the economic or health consequences is utilitarian and applying moral rules or dietary regulations (e.g., kosher restriction in the Hebrew Bible) is deontological; but the Buddhist view of karma as a "virtue theory" provides a completely different perspective.

According to this interpretation of Buddhism, the most important point about karma is not whether it is a moral law involving some precise and inevitable calculus of cause and effect. The basic idea is simply that our actions have effects—more precisely, that our morally relevant actions have morally relevant effects that go beyond their utilitarian consequences. In the popular Buddhist understanding, the law of karma and rebirth is a way to get some control over what the world does to us, but this misses the main significance of the Buddha's reinterpretation. Karma, I suggest, is better understood as the key to spiritual development: how our life situation can be transformed by transforming the motivations of our actions right now. When we add the Buddhist teaching about *anatta* (nonself)—the claim, consistent with modern psychology, that one's sense of self is a mental construct—we can see that karma is not something I have; it is what "I" *am*, and what I am changes according to my conscious choices. "I" (re)construct myself by what "I" intentionally do, because "my" sense of self is a precipitate of my habitual ways of thinking, feeling, and acting. Just as my body is composed of the food I eat, so my character is composed of my conscious choices and constructed by my consistent, repeated mental attitudes. This means that we are "punished" not for our sins but by them. From the other side, Spinoza expressed it well: happiness is not the reward for virtue but virtue itself. People bear the consequences not for what they have done but for what they have become, and what we intentionally do is what makes us what we are. An anonymous verse sums this up quite well:

> Sow a thought and reap a deed
> Sow a deed and reap a habit
> Sow a habit and reap a character
> Sow a character and reap a destiny

But what sorts of thoughts and deeds should be sown? One of the ways that Buddhism explains our *dukkha* is by tracing it back to the "three poisons" (also known as the "three unwholesome roots," or *akusala-mula*) of human motivation: greed, ill will, and delusion.[20] To reduce our *dukkha*, these need to be transformed into their positive counterparts: greed into generosity, ill will into friendliness, and the delusion of a separate self into the wisdom that acknowledges our nonduality with the world. Such an understanding of karma does not necessarily involve another life after we physically die. To become a different kind of person is to experience the world in a different way. When my mind changes, the world changes for me, and when I respond differently to the world, the world responds differently to me. Since we are not separate from the world, our ways of acting in it tend to involve reinforcing feedback systems that incorporate other people. The more I am motivated by greed, ill will, and delusion, the more I must manipulate the world to get what I want, and consequently the more alienated I feel and the more alienated others feel when they realize they have been manipulated. This mutual distrust encourages both sides to manipulate more. Moreover, the more my actions are motivated by generosity, loving-kindness, and the wisdom of nonduality, the more I can relax and open up to the world. The more I feel part of the world and at one with others, the less I am inclined to use others and, consequently, the more inclined they will be to trust and open up to me. In such ways, transforming my own motivations not only transforms my own life; it also tends to affect the motivations of those around me.

This may or may not be interesting, but what does it have to do with genetically modified food? Before we can apply these teachings in that direction, it is necessary to say a little more about the Buddhist solution to delusion, which means emphasizing the "three basic facts" that have already been mentioned in passing: *dukkha* "dis-ease," *anicca* "impermanence," and *anatta* "nonself."[21]

Dukkha is the most important term in Buddhism, for the Buddha emphasized that his only concern was showing how to end it. This suggests that Buddhism is a form of utilitarian consequentialism, but, as the above explanation of karma implies, Buddhism has a rather subtle understanding of what makes us unhappy—more precisely, of how we make ourselves unhappy. *Dukkha* is usually translated as "suffering," but *dukkha* encompasses much more than what we normally understand as suffering because it is intimately related to the other two basic teachings, impermanence and nonself.

Anicca means that nothing is eternal, everything arises and passes away according to conditions, including ourselves. Socially, this also implies an openness to change, including progress—if it really is progress, that is, an improvement of previous conditions. New technologies are not in themselves a problem, for the important issue is their effect on our *dukkha*. Buddhism is not nostalgic for some prelapsarian age in the past when life was "natural" because there never was such a golden age.

Anatta (nonself) was already mentioned in the discussion of karma, yet it has several other implications important for a Buddhist evaluation of genetic modification. The delusion that *anatta* resolves is duality: our usual sense of self as something separate from the external world that we are "in." This is the most troublesome *dukkha* of all because the groundlessness of our constructed self is usually experienced as a sense of lack that we are unable to resolve: we feel the need to make ourselves more real, more substantial, but we can never do so because the self is a psychological and social construct having no "substance" of its own that could become real.

In contrast, the wisdom of *anatta* is realizing that nothing has any "self-essence," not only because there is no permanence, but also because everything is interdependent on everything else, part of a web so tightly woven that each phenomenon in the universe is both effect and cause of all other phenomena. This "interpermeation" of everything is well expressed by Thich Nhat Hanh, a Vietnamese teacher (and poet):

> If you are a poet, you will see clearly that there is a cloud floating in this sheet of paper. Without a cloud, there will be no rain; without rain, the trees cannot grow, and without trees we cannot make paper. The cloud is essential for the paper to exist. If the cloud is not here, the sheet of paper cannot be here either. . . .
>
> If we look into this sheet of paper even more deeply, we can see the sunshine in it. If the sunshine is not there, the tree cannot grow. In fact, nothing can grow. Even we cannot grow without sunshine. And so, we know that the sunshine is also in this sheet of paper. The paper and the sunshine inter-are. And if we continue to look, we can see the logger who cut the tree

and brought it to the mill to be transformed into paper. And we
see the wheat. We know that the logger cannot exist without
his daily bread, and therefore the wheat that became his bread
is also in this sheet of paper. And the logger's father and mother
are in it too. . . .

You cannot point out one thing that is not here—time,
space, the earth, the rain, the minerals in the soil, the sunshine,
the cloud, the river, the heat. Everything co-exists with this
sheet of paper. . . . As thin as this sheet of paper is, it contains
everything in the universe in it.[22]

Notice that this way of deconstructing substantiality or "separate
thing-ness" (Sanskrit, *sva-bhava*, literally "self-being") does not discriminate
between natural phenomena (sun, rain, trees) and more technological ones
(e.g., the chainsaw that the logger uses or the paper mill that processes the
wood pulp).

In sum, nothing has any reality of its own, because everything is part of
everything else. Nevertheless, that we do not need to worry about disturbing
genetic "essences" does not liberate us to do whatever we want technologically. The most important criterion for Buddhism remains *dukkha*: does a
genetic modification tend to reduce that or increase it? Emphasis on the
interdependence of everything complicates such an evaluation, and, unsur-
prisingly, this is where there have been the most problems, due to unexpected
"side-effects": Bt corn pollen alleged to kill monarch butterflies, allergic reactions due to the Cry9C protein in StarLink corn, cross-pollination of native
Mexican maize with GM varieties, genes from genetically engineered plants
unexpectedly ending up in the guts of honeybees, and so on.

Samuel Abraham's earlier chapter in this volume on the science of
genetic engineering emphasizes how the expression of each gene in a genome
is regulated by the proteins expressed by other genes. "By the extension of
this scenario virtually all the genes in an organism are now inextricably
linked to each other's expression." Whether genetic modification "will cause
a measurable change to the organism's full range of functions and responses"
would seem to be a central question, according to Abraham, yet evaluations
of GM food normally test only for their safety (toxicity and allergenicity)
and comparative nutrient value. The guidelines used in such evaluations
"have a very utilitarian human-centric perspective with the plants only considered in the context of how well they suit our needs."

What do all these Buddhist principles imply about GM food? It is easy
enough to see how the three unwholesome roots of motivation—greed, ill
will, and delusion—function on the personal, individual level. Given the
focus of this book, a crucial question is whether they also operate collectively

and institutionally—in the food industry, perhaps? *Can the three unwholesome roots of motivation be institutionalized?* If so, does the law of karma still imply unwholesome results? Not in some magical way, but in the sense that we reap what we sow? In other words, where do the motivations behind the development and introduction of GM food fit into the collective karmic process, as understood by Buddhism?

For a brief period, "golden rice" genetically engineered to include beta-carotene (which our bodies convert into vitamin A) was proposed for nutritional deficiencies in some undeveloped countries, until it was realized that the amount of beta-carotene that could be added was too small to be significant.[23] A more significant and notorious example of GM, however, was Monsanto's attempt to introduce a patented "terminator gene" into the world's main food crops, which it gave up only because of very damaging publicity. I am inclined to emphasize that case more because the genetic modifications I have read about seem designed to help the food industry more than the food consumer. The focus is on growing and processing food more efficiently, rather than on taste or nutrition. In a competitive industry, this may end up reducing consumer prices, yet it is not otherwise clear to me that GM in food actually working to reduce consumer *dukkha*.

On the other side, what has already occurred, repeatedly, are unexpected problems, usually for those who have not asked for GM food and perhaps have little to gain from it. A few examples were cited earlier. Monarch butterflies feed exclusively on milkweed leaves, and in 1998 it was claimed that milkweed contamination from Bt corn pollen was killing them. Also in 1998, Arpad Pusztai, a scientist working in Britain, reported that in his experiments genetically modified potatoes were causing immune system damage to rats. In 2000 StarLink corn, with a protein indigestible to humans, was accidentally released into the human food chain, leading to thirty-seven reports of serious allergic reactions investigated by the U.S. Food and Drug Administration. In 2001 Ignacio Chapela and David Quist, researchers from the University of California, claimed to have discovered that genes from biotech corn had contaminated native maize in the Mexican highlands,[24] and so forth.

One more incident is worth emphasizing. In May 2000 Monsanto revealed some amazing new information about its Roundup Ready GM soybeans. Four years after they began to be consumed, the company told the USDA that it had just discovered two "unexpected" DNA fragments in its genome, one 250 base pairs in length, the other 72 base pairs, which had somehow been unintentionally inserted—or else had been there all along, but without the company's knowledge of their presence.[25]

There are at least two reasons to be concern about these incidents, in addition to the specific problems they reveal. First, they suggest what the

Buddhist emphasis on interdependence also implies: that meddling with the genome of food plants (and no doubt that of animals as well) is an extraordinarily complicated process with many types of subtle consequences that are very difficult—perhaps almost impossible—to anticipate and evaluate exhaustively. In other words, we can expect these types of accidents to recur indefinitely. Second, almost as disturbing has been the reaction of the food industry, which has tried to deny these incidents, minimize them, and—particularly in the cases of Pusztai, and Chapela and Quist—has undertaken questionable public relations campaigns to impugn their scientific competence and personal integrity.

What do these concerns reveal about institutional motivation? We are reminded that the food industry is a food *industry*. Inevitably, then, providing nutritious and healthy food is not the ultimate *goal* in this system but the *means* within a larger economic process in which the focus, naturally, is profitability and efficiency. For our economic system, food is another commodity. Genetic modification does not make food into a commodity—it is already a commodity—but the safety problems with GM food make us more aware of the problems with commodifying food, because producing safe and nutritious food appears to be more complicated than providing most other consumer products. The important question becomes: given the extraordinary difficulties with testing for possible adverse effects, along with corporate pressures for short-term profitability and growth, can the food industry be trusted to subordinate its own interests in GM and place top priority on safeguarding the needs, not only of human consumers, but of the whole ecosystem? Furthermore, given the strong corporate influence on many governments—especially in the United States—can the Food and Drug Administration be trusted to prioritize the needs of consumers and the biosphere?

In terms of the karmically-significant motivations discussed earlier, the larger ethical problem here might well be characterized as *institutionalized greed*.

Does this go along with *institutionalized delusion*? If there is a problematical duality between the institutional interests of food producers and individual interests of consumers, there is a much greater one between the human species and the rest of the biosphere. Since the advent of the modern era, our escalating technological powers have been used to subdue and exploit the rest of the biosphere, with little concern for the consequences of our domination for other species. We continue to act as if we have no responsibility for the other beings with which we share the earth, as if they have no value or meaning except insofar as they serve our purposes. One consequence of this duality is the extraordinary biological success of the human species, by any measure; but the underside of this evolutionary success is an ecological crisis that is already seriously affecting the quality of our own lives. The nonduality between us and the rest of the biosphere—a collective version of the wisdom that Buddhism emphasizes—means that these two consequences are inseparable. There are no "side-effects," only consequences we like and consequences we do not like. Since we are part of the natural world, if we make nature sick, we become sick. If the biosphere dies, we die. This is as good an example of karma as we will find.

What does this imply about the karma of GM food? The genetic modification of food is only a small part of this larger commodification process, but a significant part of it, since technological modification of other plant and animal species, without a much better understanding of their genomes and how all the genomes of living creatures affect each other, is an especially dangerous example of how our ambitions tend to outrun our wisdom. In this way, I am led to conclude that genetic engineering of food, as presently practiced, is probably incompatible with basic Buddhist teachings, because inconsistent with the kinds of personal and collective transformation of motivations necessary if *dukkha*—not only human *dukkha* but that of other living beings too—is to be reduced.

This does not necessarily mean that the genetic modification of food is always a bad thing to be avoided, which would be an "essentialist" claim inconsistent with the primary Buddhist emphasis on reducing *dukkha*. Since Buddhism does not privilege "the natural," including the natural selection that drives the evolutionary process, there is the possibility that in the future some modifications might actually serve to reduce *dukkha*. Although it would need to be very carefully tested, there is the possibility that vitamin A–enriched rice might someday be a benefit to humankind without being a threat to the rest of the biosphere.

From a Buddhist point of view, technologies are neither good nor bad in themselves. Nor are they neutral. That is because technologies cannot be separated from the larger social, economic, and ecological contexts within which they are devised and applied. The Buddhist understanding of karma as *cetana* implies that personal and institutional motivations are an essential part of that context. This means that any attempt at evaluating a technology such as the genetic modification of food needs to take the motivations behind those innovations into account.

Notes

1. Today almost all Buddhist monastics are *bhikkhu* monks; except in Taiwan the *bhikkhuni* order of nuns has largely disappeared, although there are efforts to revive it.

2. See Vinaya Pitaka IV.888, 23–26.

3. Literally, "pig's delight," which some commentators claim might have been a mushroom poisonous to humans.

4. Mahavamsa of the Vinaya Pitaka VI.23, 10–15. The Theravada Buddhist focus group mentioned the "Big Four": tiger, lion, horse, and elephant.

5. John Kieschnick, "Buddhist Vegetarianism in China," in Roel Sterekx, ed., *Of Tripod and Palate* (London: Palgrave Macmillan, 2005), p. 189. I am grateful to Kieschnick for making his paper available to me before publication.

6. Kieschnick, "Buddhist Vegetarianism."

7. See especially the Lankavatara Sutra, trans. D. T. Suzuki (London: Routledge Kegan Paul, 1932), pp. 211–23.

8. Japanese Buddhist priests, who commonly marry, drink alcohol, and wear lay clothes, are rarely vegetarian, although meat and fish are not usually consumed in monasteries.

9. Kieschnick, "Buddhist Vegetarianism," p. 194.

10. According to the Theravada Vinaya Pitaka IV.259, 15), only garlic (believed to be a sexual stimulant) is forbidden for *bhikkhu* and *bhikkhuni*, although allowed when one is ill.

11. The Sarvastivada Vinaya Pitaka gives the story of monk who did not attend a sermon of the Buddha because he had just eaten garlic and did not want to offend others; the Buddha scolds him for preferring garlic to the Dharma and establishes the rule that monks should not eat garlic unless they are ill. (Kieschnick, "Buddhist Vegetarianism" p. 190)

12. Retrieved on 12 June 2005 from http://www.gene.ch/gentech/1997/8.96–5.97/msg00050.html and http://www.netlink.de/gen/Zeitung/1221.htm. The Dharma Realm Buddhist Association is a large (26 branch monasteries and temples) Buddhist lineage based in California and founded by Hsuan Hua (1918–95), a Chinese Chan master.

13. A possible exception is addressed later.

14. Retrieved 12 June 2005 from http://www.camradvocacy.org/cell/news.asp?id=847. My italics.

15. The only Buddhist term that might be understood as referring to "natural law" is *dhamma* (Sanskrit, dharma). It has three primary meanings: the teachings of Buddhism; "the way things are"; and (more technically) the basic elements of reality, according to the atomic ontology of early Buddhist abhidhamma philosophy. The second meaning comes closest to a Western view of nature, but it is usually explained with the three "facts of existence" to be discussed later.

16. Lambert Schmithausen, "The Early Buddhist Tradition and Ecological Ethics," *The Journal of Buddhist Ethics* 4 (1997). Retrieved on 29 August 2005 from http://jbe.gold.ac.uk/4/schm1.html.

17. The five basic moral rules to be followed by all Buddhists: to avoid killing living beings, stealing, lying, improper sex, and alcohol. The first precept suggests that it may be wrong to genetically engineer plants to kill insects (e.g., Bt corn), but this issue was not raised in the Theravada Buddhist focus group or the Chinese traditions focus group.

18. This is already the case for most turkeys, although as the result of selective breeding rather than genetic modification. From the Buddhist perspective, of course, their poor quality of life is no less objectionable.

19. Anguttara Nikaya 6.63, in Nyanaponika Thera and Bhikkhu Bodhi, trans. and ed., *Numerical Discourses of the Buddha: An Anthology of Suttas from the Anguttara Nikaya* (New York: Altamira, 1999), p. 173. The Pali term kamma (Sanskrit, karma) literally means "action," while the "fruit" of such action is *phala*, but in popular Buddhism the meanings of the two terms are often conflated.

20. "There are, O monks, three causes for the origination of action. What three? Lobha greed, dosa hatred and moha delusion." (Anguttara Nikaya III, in *Numerical Discourses*, pp. 49–50.)

21. Anguttara Nikaya III, in *Numerical Discourses*, p. 77. The Theravada Buddhist focus ggroup mentioned these as the "three fundamental things that the Buddha taught."

22. Thich Nhat Hanh, *The Heart of Understanding* (Berkeley: Parallax, 1988), pp. 3–5.

23. As I write, there are new attempts to genetically increase the amount of beta-carotene in GM rice.

24. These examples are from Kathleen Hart, *Eating in the Dark: America's Experiment with Genetically Engineered Food* (New York: Vintage, 2003), pp. 138–46, 47–53, 229, 280. All of these claims have been contested and remain controversial. For example, some more recent studies have not been able to confirm the GM contamination of Mexican maize.

25. Ibid. 219.

9

"So That You May Have It with No Harm"

Changing Attitudes toward
Food in Late Imperial China

Hsiung Ping-chen

Eating Habits of One-Quarter of Humanity

At this point, few should quibble about the importance of learning more about China's food culture and the eating habits of a billion or more people. In terms of raw numbers and sheer market economics founded on demographic statistics, we are looking at a collective appetite that represents roughly one-quarter of humanity,[1] famous for their excessive interest in food.[2]

For an agrarian society that traditionally relied on rice and other grains for its staple food, this chapter looks at changing attitudes toward rice and grains to reflect on food and dietary habits in late imperial China. It begins by surveying the cultural history of food and diet in early China. Then it discusses foodstuffs as medicinal devices, on the one hand, and as daily diet, on the other. Third, specialized recipe booklets from the eighteenth and nineteenth centuries are examined to show the case of congee as a typical food choice that represents the characteristic of this gastronomical tradition. In so doing, it is also shown how considerations of palatable taste and physical survival surfaced as the ultimate criterion, and not the market economics of agrotechnology, in the handling of rice and grains as important foodstuffs during this period. As the slogan of the day had it, the secret lies in the idea that as far as any foodstuff goes, it should always allow people "to have it often, in quantity, and over a long period of time without being harmed"—a guideline that may still be of use when considering genetically manipulated foods in modern society.

Historically, in what is now referred to as "Chinese food culture" there is reflected a long evolutionary process in the East Asian continent. As a culinary art, like many other world food cultures, it would certainly be more apt to speak of it as a living representation, a matrix, of various dietary traditions rather than a single "eating" habit. Nevertheless, for many who have studied it, Chinese food culture does exhibit certain distinct features and a detectable continuity, as much as a wide variety and regional characteristics.[3]

During this long evolutionary process, one peculiar feature has shone through, especially in a comparative perspective. Dietary preference and avoidance of certain foods, though certainly known to this population, were never at the center of their daily practice. The Chinese attitude toward and practical use of food have instead been founded upon useful, if intuited, principles and sound values rather than upon the enunciation of explicit taboos or clear prohibitions. Early on, Taoism, or for that matter Buddhism and the Chinese popular religion that came after, rarely proscribed eating habits as might have been the case elsewhere. The foods people have, in this broad Sinic world, reflect primarily the decisions people make in selection of foodstuffs, cooking styles, and consumption.[4] Such food-related preferences may be rooted in ethical and religious backgrounds, yet they are not food taboos derived directly from a belief system.

Part of the background for this relatively relaxed attitude toward food prohibitions comes from a belief that the Chinese body was itself a microcosm. According to Taoist beliefs, human physiology functioned as a system of contrasting and mutually complementary forces. Familiar factors such as the yin-yang, or the five elements (metal, wood, water, fire, and earth) worked against but also complemented one another. For instance, in terms of gastronomical ingredients, all food properties may be grouped in the two categories of yin and yang: yin (i.e., cool, or *liang* 涼) foods include many leafy vegetables, watery plants, and summer fruits, whereas yang (i.e., hot, or *je* 熱) foods include fatty meats, oily nuts, gingery or peppery spices, and alcohol. The point was not to fight over which item was more beneficial or damaging than another, as much as finding a perfect balance.[5] It was hoped that, eventually, a harmony would be struck and stability achieved while discord and problems would be avoided.[6] On further analysis, yin was identified as representing the dark, cold, or feminine force and yang the bright, hot, or masculine force. In addition, since people varied in their physical dispositions to begin with, they might be inclined toward either the yin or the yang, thus needing the counterbalancing force of the opposite.[7] Furthermore, understood not merely as a source of nutrients mechanically providing for people's physical survival, food, when chosen and taken wisely, was believed to bring about and enhance health. Thus for all ends and purposes, food was always "the way of maintaining bodily harmony," a tangible way to achieve material balance.[8] In other words, if people ate correctly, life was nurtured, and their health was strengthened because some fundamental cosmological principles were being observed.[9]

Culinary preparation too could produce either a cooling or warming effect. Cold water infusion through boiling and, on the other hand, deep-frying, roasting, or baking, are examples of the two extremes of a wide spectrum, where the increasingly popular method of stir-frying represents an interesting medium. For the ordinary consumer, therefore, in order to modify physical disposition and maintain an equilibrium, careful selection of food materials and cooking styles was critical. For the longer a foodstuff was cooked (e.g., in a stew or dry roast), the more heat it gathered.[10] And for an already-balanced body, over-consumption of a certain food or an unwise choice of culinary options could well tip the balance.[11]

In matters of dietary ingredients and culinary preparation, common consumers were therefore the ultimate arbiters. In the taking of one's daily meals, for example, to enhance the equilibrium and to achieve health, there had always to be soupy liquids (*yin* 飲) to go with the drier food (*shih* 食). Likewise, on a table, there had to be supplementary dishes (*fu-shih* 副食) to complement the main staple (*chu-shih* 主食) according to regional options. In the south, this was known as the principle of *fan* (飯, cooked rice or grain) and *ts'ai* (菜, complementary dishes).[12] For ordinary folks, at the end of the day, the idea was never to allow one's greed to take over prudence in consumption. Frugality dictated "seventy percent full (*ch'i-fen pao* 七分飽)" as the golden rule for people taking their meals.[13]

Food, Health, and Medicine

One of the long-held practices within the Chinese food tradition was to intentionally blur the boundary between food and medicine, such that medicinal effects were commonly expected from ordinary food intake, and therapeutic prescriptions might also assume the form of common foodstuffs in culinary preparations. To regularly mix the two categories into one was never considered awkward or backward. Quite on the contrary, it was deemed a marvelous achievement, a creative blend that made medicine more acceptable while allowing common foods to take on additional "magical properties."[14]

In this regard, food became medicine because in addition to its curative properties, the daily consumption of food, when properly selected and thoughtfully handled, was believed to have the property for preventing the advent of disease, thus leading to a positively healthier life.[15] This tradition of adopting food as medication and integrating food with therapeutic treatments dates back to at least the Chou dynasty (1066–256 BC).[16] In Chou palaces, culinary experts were always available to attend to the health needs of royal dignitaries, and the yin-yang principle and the five elements acted as the basic reference.[17] In this schematization, food was divided into different subgroups corresponding to the yin and the yang categories, so that taste was formulated according to the five-element principle. Bitterness, saltiness, sourness, spiciness, and sweetness were thus coordinated with the five elements respectively so that gastronomical as well as pharmaceutical properties were constructively maintained.[18]

The Case for Congee

Many features of Ts'ao T'ing-tung's (曹庭棟 1699–1785) creation, *The Book of Congee* (*Chou-p'u* 粥譜) from the eighteenth century, touch upon the fundamental principle of food intake in the Chinese tradition and warrant a close examination as a case par excellence representing basic concerns reworked at a critical juncture in Chinese culinary history.[19] In the short preface, "On the Book of Congee (Chou-p'u shuo 粥譜說)," preceding the one hundred congees that make up the main body of *The Book of Congee*, the author points out that since congee had long attracted both the medical practitioners and folk experts in pursuit of longevity,[20] various congee recipes were to be found in prescription booklets.[21] With such opening lines, the specific point that Ts'ao T'ing-tung, who called himself the "Gentleman from the Mount of Compassion (Ts'u-shang chü-shih 慈山居士)," wanted to emphasize was, "According to my own humble opinion, congee is really for daily use and ordinary consumption (*jih-yung ch'ang-kung* 日用常供)."[22] In other words, understanding that most of his audience was of the opinion that the thin line between food and medicine should be left at that, he chose to place his emphasis on the appreciation of foodstuffs rather than pharmaceuticals. This argument was that even though these grains loosely cooked together with various compounds could be recognized as "occasionally of use in the treatment of illnesses (*ou tzu chih chi* 偶資治疾)," the pertinent issue here was that the thick, soupy preparations fell under the name of congee (*chou* 粥) as comfort foods; the issue of taste therefore became of critical importance.[23]

To lay his argument, Ts'ao first makes clear his conviction that congee, as a form of porridge and also an aliment especially beneficial to health, has had a long history in serving the needs of the poor, the sick, the very young, and the elderly. He also admits that as a heritage well-known to and long practiced by Taoist believers and people pursuing the art of longevity, the myriad of recipes flooding the market, with their wide variety in ingredients and cooking styles, tantalized their daily consumers with perhaps too large an array of tastes. And that, Ts'ao says, motivated him to put matters in order according to his own thinking. The idea is, first of all, to work against the old habit of seeing congee "as a medicine" and instead promote it steadfastly "as a meal in itself and a source of daily consumption."[24] Because congees boil and simmer slowly, they borrow various properties from different sources to produce fortified food stuff as a thick, wholesome meal in itself. They thereby embody the Chinese dietary principles of balance, subtlety, and modesty, as well as the idea of complimentarity, which made consumers and promoters marvel over them as unique culinary creations. For admirers such as Ts'ao, one way to drive home this point was to separate congee's common character more clearly from its more weighty importance in medicine. The key for a gastronomical expert like Ts'ao was to lay a new emphasis on congees' palatable taste and aesthetical quality rather than whatever spectacular properties or physiological implications they might be said to possess. In other words, he liked to direct his deliberation on porridge away from its efficacy as a remedy while inducing ordinary consumers to concentrate on the comfort and taste afforded in a meal. The device he chose to demonstrate this principle was to regroup congees into three categories of superior (shang-pin上品), average (chung-pin中品), and lowest (hsia-pin下品) types, based on the sole consideration of taste: "Regardless of their curative effect, the best congees, classified as superior, are those that are lightly fragrant and palatable; next stand the average which are not as sweet-scented or tasteful; whereas included in the lowest category are those with a strong odor and turbid appearance."[25] Altogether, one hundred items are thus categorized, with the most disagreeable of the recipes completely purged from his list. These best-known and now well-ranked instructions, Ts'ao insists, have been collected from previous gastronomical authorities, with due references to show respect and accord credibility.[26] In addition, all recipes were properly tested to ascertain their practicality. For any with medicinal properties, their pharmaceutical functions were also observed, along with the reminder that while prescriptions may appear clear, their curative function could hardly be fixed, thus ought to remain flexible.[27]

The Importance of Rice, Water, Fire, and Consumption

For the culinary preparation of congee Ts'ao selected the four factors of rice, water, fire (or heat), and ways of consumption as being of primary concern. To each he devoted a separate deliberative essay, at the end of which, he penned in a few sample recipes other than the easily accessible, all with demonstrable benefit to the palate and people's health, especially in assisting the needs of the old and the frail. Given the historical (and medicinal) background of various congees, Ts'ao was fully aware of the implications of his attempt to change people's ideas about congee. Thus, at the end he recalls that the only rule regarding food preparation and consumption has always been that it be beneficent (toward people's needs) and palatable (to their taste) and asks rhetorically: according to these principles, "could there be anything that would prevent one from creating a tradition of one's own (tzu-wo tso-ku 自我作古)?"[28]

First, on "Selecting of the Rice (tse-mi 擇米)," Ts'ao makes it plain that though a wide variety of rice could be used to prepare congee, they all represent a distinct choice: short-grain nonglutinous rice, for instance, was the best and especially fragrant. In terms of the season, late-harvested rice was also quite acceptable for its "soft" character, whereas the early-harvested kind was not as desirable. Furthermore, since stored rice was not as "rich and smooth," while the grain newly husked in autumn had a rich aroma, people should husk it only when needed and store it in a well-ventilated place. There were also those who liked to cook with roasted rice (ch'ao pai-mi 炒白米) or slightly burned rice crust (chiao kuo-pa 焦鍋芭), for their hot and fragrant aroma was supposed to expel dampness while producing an appetizing effect, even though the taste was not as rich and smooth. As stated in the Compendium of Materia Medica (pen-ts'ao kang-mu 本草綱目), congees made with short or long-grain nonglutinous rice, millet, or sorghum were known to be diuretic, thirst-quenching, and nurturing for the spleen and stomach, whereas those made with glutinous rice or other grains (sticky sorghum, broomcorn millet, etc.) could cure people of asthenia-coldness, diarrhea, and vomiting.[29] Knowing that medicinal congee recipes rooted in pharmaceutical prescriptions had an illustrious past record, Ts'ao was of the opinion that congees prepared with rice and other grains should be treated as ordinary food for daily consumption. As such, the rice or other grains as a plain but key ingredient might still help to mitigate the excessively potent elements in the recipes or enhance those with an intrinsically mild character. It was such balancing effects that revealed the truly "magical" power of congee as an attractive food form.[30]

After the rice or other grains had been chosen as the main ingredient, next came consideration of the water or other fluids used in congee making. According to Ts'ao's advice, water of different provenance was never of uniform properties. Inappropriate methods of procurement and boiling were therefore bound to do damage to the taste of a congee. Good sources of water one should choose to cook with included early spring rain (for it embodies the vital energy of flowering yang), melted winter snow (antipyretic and good for treating seasonal diseases), fresh running water in all seasons, and well water (especially that from the first draw of the morning). Problematic waters, however, included plum rains (*mei-yü* 梅雨) (falling during too hot and humid periods, which made mold grow and people fall sick), summer or autumn rain (often after a deluge, therefore unsuitable to cook with), melted spring snow (easily spoiled and susceptible to worm larva), mountain spring water (whose quality changed according to locality), and water from ponds or swamps (often poisonous). Congees prepared with water of the finest kind, such as the lightly greenish well water from the morning's first draw, guaranteed the culinary creation a natural sweetness. Ts'ao claimed this was the heavenly *ch'i* (氣) surfacing from the fluid, and he insisted that nothing else was needed for the congee preparation: the gastronomic result, no doubt, would be simply extraordinary.[31] In addition, Ts'ao told people to store usable water in a vat with a piece of cinnabar (*chu-sha* 硃砂) at the bottom to dissolve various poisons (*pai-tu* 百毒), thus giving people the longevity they sought.[32]

Third, concerns for the cooking fire and the use of tinder woods were also of crucial importance in the preparation of congee. Here Ts'ao thought the key had to do with the kind of heat the fire produced. Since, when making congee, the principle was that it be cooked until a viscous quality was achieved, excessive or inadequate heat could affect the final taste, rendering it too thick or unacceptably thin. In this regard, mulberry kindling was considered the top choice by classic authorities. Oak was not as good, for it produced too intense a heat, although for those congees that required constant boiling, oak tinder could be useful.[33] In congee-making, Ts'ao insisted that the water first be boiled to remove impurities. While boiling, it should also be ladled off with a spoon before grain or rice was added. The boiler should preferably be made of porcelain, not metal.[34] These deliberations on choice of container, tinder fire, and grain all illustrate his point that the preparation of congee constituted a gastronomical whole, of which no part could be neglected without compromising the final product. Ts'ao's last piece of advice against common mistakes in congee cooking warned people against the habit of placing porcelain cookers directly into the furnace (or under the stove) to make the liquid simmer away while burning off rice stalks or grain husks. This method, though seemingly convenient, he said, was in fact unadvisable because the strength of the fire could not be effectively monitored.[35]

Fourth, in the essay about the moment for "consumption (*shih-hou* 食候)," Ts'ao stressed the understanding that the taste, key to the enjoyment of congee, lay in the way it was consumed. Older people, Ts'ao conceded, might well choose to partake of it at any hour of the day they pleased: they had arrived at an age when they no longer had to worry about fixed mealtimes

but could ingest food whenever they felt hungry, aiming at achieving a
strong body while enjoying a long life. For the daily nurturing of the ordinarily healthy common folks, however, congees were best taken on an empty
stomach or as a supper. The point was that nothing else should be eaten
afterward. The other principle was that people ought not eat themselves too
full of congee. It was true, Ts'ao admitted, that this being the kind of food
that will not stay long in the stomach, there should be little apprehension
about overstuffing. Still, the slightest feeling of fullness could cause damage
to the stomach and should be avoided.[36] Another caution was that people
should rather have it slightly overheated. Even though a piping hot congee
might cause people to sweat a little, that also helped with perspiration and
circulation of the blood.[37] Finally, people should try not to consume congee
with side dishes or elaborate condiments, for fear that the body might not
concentrate on absorbing the congee as the main nutrient. At most (if
necessary), one might only brush the lips with some salt to add flavor if
the congee alone tasted too bland.

The Taxonomy of Congee and the Ranking of Foodstuff

The way Ts'ao ranked congees, one hundred in total, into three distinct
groups reveals most about his culinary prejudices about food. For in the
thirty-six that he chose as belonging to the superior category (shang-pin 上
区), only one contained any meat product (that of a congee prepared with
dried deer's tail).[38] All the rest were made with beans, nuts (almonds, pea-
nuts, sesame seeds), flower buds (lotus, chrysanthemum, or plum), or other
fragrant herbs (mints or teas).[39] The congee was to be light in color and
fragrance, light in idea and consumption, and understated in taste. Congee
prepared with mung beans (lü-tou chou 綠豆粥), coix seeds (i-i chou 薏苡
粥), or lily bulbs (pai-ho chou 百合粥) may be included here as the best
examples of the kind of recipes selected according to this consideration.
Congee made with ginger and red dates (chiang chou 薑粥) was another good
example in this first rank, popular with the ordinary people and producing
well-tested health benefits.[40]

Included in Ts'ao's average group were twenty-seven congees, still
mostly made with vegetarian ingredients. Although those prepared with
delicately flavored animal products such as cow's milk (niu-ju chou 牛乳
粥), chicken broth (chi-chih chou 雞汁粥), and duck broth (ya-chih chou 鴨
汁粥) were added to this rank.[41] Congees with venison (lu-jou chou 鹿肉
粥) and mussels (tan-ts'ai chou 淡菜粥) stood out as the two items prepared
with meat or seafood favored by people of the lower Yangtze Valley.[42] For
the majority of recipes in this group, however, these were congees cooked

with fruits, roots, beans, or other vegetable ingredients that appeared warmly inviting but perhaps not of the most elegant character, as represented by the first rank. Congees made with longan aril (*long-yan-jou chou* 龍眼肉粥), white China root (*pai fu-ling chou* 白茯苓粥), red bean or phaseolus seeds (*ch'ih hsiao-tou chou* 赤小豆粥), or mountain yam (*shan-yao chou* 山藥粥) are good examples of the kind of selections that went into Ts'ao's selection for this middle-ranking group.[43]

For the thirty-seven congees that Ts'ao ranked as least favorable, the criteria are clear.[44] Here could be included the popular (in his opinion different from "enjoyable") sorts, like radish or spinach congee; those endowed with proven medicinal efficacy, like rehmannia root congee (*ti-huang chou* 地黃粥) or ophiopogon root congee (*mai-men-tung chou* 麥門冬粥); or those cooked with precious ingredients (*lu-sheng chou* 鹿腎粥, deer kidney congee, *yang-sheng chou* 羊腎粥, goat kidney congee). The point was that their tastes were invariably strong.[45] In this respect they were like other kinds of congee imbued with a strong taste (such as that made with sour dates, *suan tsao-jen chou* 酸棗仁粥), strong color, heavy meaty taste (*chiu-yeh chou* 韭葉粥, chive congee), or rare and precious (but to him, gastronomically repulsive) ingredients, such as deer or lamb kidneys, or pigs' backbone.[46] Those congees cooked with dog meat (*chuan-jou chou* 犬肉粥), mutton (*yang-jou chou* 羊肉粥), evodia fruit (*wu-chu-yü chou* 吳茱萸粥), or magnetite (*tz'u-shih chou* 慈石粥) give us a crude idea as to the kind of congees cast into this lesser category.[47]

The Need for Subsistence versus the Pleasure of Eating

In the second half of the nineteenth century, another gentleman-scholar from the middle Yangtze Valley, Huang Yün-hu (黃雲鵠1819–1898), produced two booklets related to congee and congee consumption that bear a different value from Ts'ao's, because both Huang's works emphasized the critical importance of the availability rather than the taste, in other words valuing economics over aesthetics, of food pertaining to congee.

In his preface to its print release, also titled *The Book of Congee* (Chou-p'u 粥譜), Huang stresses the unique value of congee as a "poor folks' food" and famine relief.[48] Quoting the proverb that usual yield of "infertile lands could only provide people with thick porridge (*po-t'ien i kung chan-chou* 薄田以供饘粥)," he asked all to appreciate "congee" as the best food choice for the deprived.[49] For him, the question was never to focus on the enjoyment of congee for those who could afford a comfortable life but its attraction as a diet that had always fed a large number in times of want.[50] As a food form, therefore, besides the fact that congee had been praised as the best food for nurturing the elderly, the young, and the weak, easily digestible, and soothing to the respiratory system, most importantly it was inexpensive and thus easily available.[51] For a person of modest means growing up in the mid nineteenth century, when parts of China were showing signs of decline and economic crisis, Huang saw in the eating of congee a way to survive. As he wrote, he thought of his two sisters who perished young, whose "stomachs had trouble digesting congee," and of his father who had died young, leaving behind an indigent household who had to make do with habitual consumption of congee.[52] As he composed, he also remembered the comments of his old servant, who once begged to flee the smell of a food kitchen, for "[one] may not be afraid of dying but [would] dread the feeling of hunger."[53]

For Huang, the question of food was always about availability, not gratification. His second booklet, entitled *The Book on the Dispensation of Congee* (*Kuang chou p'u* 廣粥譜), was therefore written to consider the preparation of congees so that the largest number of people could be saved at times of famine.[54] He thought this had mostly to do with policy and management, never the technicalities of food preparation or a simple question of food supply.[55] Corruption and human incompetence, strategic stupidity in other words, always stood at the core of food crises, not agricultural economics.[56]

Conclusion

During the eighteenth and nineteenth centuries, the Chinese debated the subject of congee ("chou" in Mandarin or "jook" in Cantonese) as nothing but ordinary food, prepared and consumed for no other reason than people's enjoyment. In the composition and circulation, in 1773, of *The Book of Congee*, its author Ts'ao T'ing-tung, a frustrated intellectual from the lower Yangtze, proposed a thorough reform of ideas about food, using congee as his focal point. In reaction to the old love for medical efficacy, which paid little attention to the taste of or one's physical tolerance for these soupy congees, Ts'ao argued in favor of comfort and ease as well as delight and taste as the standard in thinking about food. With elaborate ideas on the kind of rice, water, tinder, and ways of consuming congee, he ranked no fewer than one hundred "designer congees" into hierarchical order. The concept was that those with a light and favorable taste became the people's first choice, while the less fine came second, and those of a heavy and thick kind, the last choice.[57]

The idea was always to employ human ingenuity in the management, handling, preparation, and consumption of foodstuffs so that they might be natural in taste and physical composition, and conform to health needs, the "spirit" being that foodstuffs had better suit the needs of the human body, not the other way around. In late imperial times, when Chinese society was at the peak of its prosperity and technological competence by premodern standards, the commentator's concern for food was not to encourage fad or invent products to benefit the market but to better delight and reward the taste and health of the common consumer: "So that everyone may enjoy food as often and as much as they wish, for a long while to come, without worrying ever about the possibility of being hurt"—that, people thought, should be the definition of good food.

Recent surveys and field reports confirm that ethnic Chinese as well as people under the influence of Chinese culture concur to a great extent in the dietary tradition and gastronomic culture uncovered here.[58] For these people, matters related to food continue to be a serious issue in their daily concerns, which makes issues related to GM food of undeniable importance. This overall concern for matters related to food leads people of Chinese descent to be apprehensive in many ways about genetic tampering with food. To begin with, they are apprehensive about the inadequate and uneven distribution of information about GM products. At a basic level, people are worried about unintentionally consuming food for lack of information, thus committing dietary mistakes, making gastronomical transgressions, due to ignorance, which compromises people's right in food choices. Like other ethnic and cultural groups studied in this volume, the Chinese have serious doubts about the human ability to modify the fundamental, natural attributes of food. Granted that, before the modern era, ample agrotechnology went into food production and food preparation in the history of Chinese science and civilization, GM food represents a categorically different step that warrants alarm. In surveys, Chinese people have expressed misgivings on the ethical and practical propriety for humans to be "playing" with food—being themselves also part of Nature. They feel humans might not have the ability to deal with the ultimate responsibility of the kind of radical alterations that some of the GM technologies imply. In addition to questions of management (e.g., who will be responsible for establishing proper regulations and due monitoring), the food-conscious are fearful of the "unknown unknowns" that GM food could represent in terms of effects on humans and the environment. To their way of thinking, it is a clear violation of nature.

Looking at the food-loving, food-sensitive, and some would say food-obsessed people that the Chinese are known to be, with their age-old beliefs in observing nature, balance, and the prolongation of life that food, when properly produced and handled, should bring, their quibbling with GM food is to be expected. That is, even though born and brought up not so much with food taboos as dietary principles and gastronomical advice, the Chinese are worried that GM food may pose a threat to their fundamental beliefs in nature, balance, and nurturing that are central to their culinary values.

As descendants with Taoist or Buddhist beliefs, the majority of Chinese people also have doubts as to the future of individual choice makers regarding food consumption. Since rice and other grains still constitute the main food items for this quarter of humanity, the history of congee studied above, with all its fablelike character, may be of relevance to the debate on GM food and related enterprises for the years ahead.

Notes

1. For a more detailed discussion of Chinese demography, see James Z. Lee and Wang Feng, *One Quarter of Humanity: Malthusian Mythology and Chinese Realities, 1700–2000* (Cambridge, MA, and London: Harvard University Press, 1999).

2. Kwang-chih Chang, ed., *Food in Chinese Culture: Anthropological and Historical Perspectives* (New Haven, CT: Yale University Press, 1977), p. 3.

3. Lu Yao-tung 逯耀東, *Tu-ta neng-jung: Chung-kuo yin-shih wen-hua san-chi* 肚大能容: 中國飲食文化散記 (Essays on Chinese dietary culture) (Taipei: Tung-ta t'u-shu ku-fen yu-hsien kung-ssu, 2001), p. 177.

4. Robert O. Whyte, *Rural Nutrition in China* (Hong Kong and New York: Oxford University Press, 1972), p. 29.

5. E. N. Anderson Jr., and Marja L. Anderson, "Modern China: South," in *Food in Chinese Culture: Anthropological and Historical Perspectives*, ed. K. C. Chang (New Haven, CT: Yale University Press, 1977), p. 367.

6. Ibid.

7. Frederick J. Simoons, *Food in China: A Cultural and Historical Inquiry* (Boca Raton: CRC, 1991), p. 268.

8. Anderson and Anderson, "Modern China: South," p. 367.

9. Wan Chien-chung 萬建中, *Yin-shih yü chung-kuo wen-hua* 飲食與中國文化 (Food and Chinese culture) (Nan-ch'ang: Chiang-hsi kuo-hsiao c-pan-she, 1994), p. 116.

10. Anderson and Anderson, "Modern China: South," pp. 368–69.

11. Kwang-chih Chang, *Food in Chinese Culture*, p. 11.

12. *Fan* was what filled one's stomach to fullness, and *ts'ai* was what garnished and highlighted a meal (whereby gastronomical experts demonstrated their lavish culinary skills). Ibid.

13. Ibid.

14. For the Chinese tradition of combining food and medicine, see Tokiko Nakayama, ed., *Chung-kuo yin-shih wen-hua* 中國飲食文化 (The Chinese dietary culture), trans. Hsü Chien-hsin 徐建新 (Beijing: Chung-kuo she-hui k'e-hsüeh ch'u-pan'she); Lin Nai-sen 林乃燊, Chung-kuo yin-shih wen-hua 中國飲食文化 (The Chinese dietary culture) (Shanghai: Shanghai jen-min ch'u-pan-she, 1989), pp. 48–52.

15. Kwang-chih Chang, *Food in Chinese Culture*, p. 9.

16. The ancient text of Chou-li (周禮, 770–476 BC) mentioned dietary medicine together with treatment of disease and trauma, healing medicine, and veterinary medicine to form the four divisions of medical practices at court. As a branch of medicine of central importance, its sphere of responsibility included the preparation of food for the royal house with the expectation that therapeutic measures be regularly implemented through dietary manipulation. "T'ien-kuan chung-tsai" 天官冢宰, in *Chou-li Cheng chu* 周禮鄭注, vol. 5, bk. 1 (Taipei: Taiwan Chung-hua shu-chü, 1965), pp. 1–4.

17. Wang Jen-hsiang 王仁湘, *Yin-shih yü chung-kuo wen-hua* 飲食與中國文化 (Food and Chinese culture) (Beijing: Jen-min ch'u-pan-she, 1993), p. 234.

18. Yao Wei-chün 姚偉鈞, *Chung-kuo yin-shih wen-hua t'an-yüan* 中國飲食文化探源 (The origin of Chinese dietary culture) (Nanking: Kuang-hsi jen-min ch'u-pan-she, 1989), pp. 7–8.

19. Ts'u-shang chü-shih 慈山居士 [Ts'ao T'ing-tung 曹庭棟], "Chou-p'u shuo" 粥譜說 (On the book for congee), in *Lao lao heng yan* 老老恆言 (The fine ways in caring for the old), vol. 5, Li-tai chung-i chen-pen chi-ch'eng 歷代中醫珍本集成, vol. 18, ed. Chu Pan-hsien 朱邦賢 and Wang Juo-shui 王若水 (1928; reprint, Shanghai: San-lien shu-chü, 1990).

20. Ibid.

21. Chia Hui-hsüan 賈蕙萱, *Chung-jih wen-hua pi-chiao yan-chiu* 中日文化比較研究 (Beijing: Beijing ta-hsüeh ch'u-pan-she, 1999), pp. 40–41, 210–11.

22. Ts'u-shang chü-shih 慈山居士, "Chou-p'u shuo" 粥譜說, p. 1.

23. Ibid.

24. Ibid., p. 1

25. Ibid.

26. Ibid.

27. Ibid.

28. Ibid.

29. "Tse-mi ti-i" 擇米第一 (Selecting of the rice), ibid.

30. Ibid.

31. "Tse-shui ti-erh" 擇水第二 (Selecting of the water), ibid., p. 2.

32. Ibid.

33. "Huo-hou ti-san" 火候第三 (Controlling the strength of the fire), ibid.

34. Ibid.

35. Ibid.

36. "Shih-hou ti-ssu" 食候第四 (Choosing the occasion for consumption), ibid., pp. 1–2.

37. Ibid., p. 2.

38. Ibid., p. 5. For a complete list of *shang-pin* congees, see Ibid., pp. 2–5.

39. Ibid., pp. 2–5.

40. Ibid., pp. 3–5.

41. Ibid., p. 7. For a complete list of *chung-pin* congees, see ibid., pp. 5–7.

42. Ibid., p. 7.

43. Ibid., pp. 5–6.

44. For a complete list of *hsia-pin* congees, see ibid., pp. 7–10.

45. Ibid., pp. 8–9.

46. Ibid., p. 9.

47. Ibid., pp. 8, 10.

48. Huang Yün-hu 黃雲鵠, "Chou-p'u hsü" 粥譜序 (Preface to *Chou-p'u*), Hsü-hsiu ssu-k'u ch'uan-shu 續修四庫全書, ed. Hsü-hsiu ssu-k'u ch'uan-shu pien-tsuan wei-yüan-hui 續修四庫全書編纂委員會, vol. 1115 (1881; reprint, Shanghai: Shanghai ku-chi ch'u-pan-she, 1997), p. 493.

49. Ibid.

50. Ibid.

51. Ibid.

52. Ibid.

53. Ibid.

54. Huang Yün-hu黃雲鵠, "Kuang chou-p'u hsü" 廣粥譜序 (Preface to *Kuang chou-p'u*); Hsü-hsiu ssu-k'u ch'uan-shu 續修四庫全書, ed. Hsü-hsiu ssu-k'u ch'uan-shu pien-tsuan wei-yüan-hui 續修四庫全書編纂委員會 vol. 1115, (1881; reprint, Shanghai: Shanghai ku-chi ch'u-pan-she, 1997), p. 503.

55. Huang Yün-hu 黃雲鵠, *Kuang-chou p'u* 廣粥譜 (The book on the dispensation of congee) Hsü-hsiu ssu-k'u ch'uan-shu 續修四庫全書, ed. Hsü-hsiu ssu-k'u ch'uan-shu pien-tsuan wei-yüan-hui 續修四庫全書編纂委員會 vol. 1115 (1881; reprint, Shanghai: Shanghai ku-chi ch'u-pan-she, 1997), p. 507.

56. Ibid., pp. 507–08.

57. Ts'u-shang chü-shih 慈山居士, "Chou-p'u shuo" 粥譜說, p. 1.

58. For the Chinese concept of food in modern times, see Vera Y. N. Hsu and Francis L. K. Hsu, "Modern China: North"; E. N. Anderson Jr., and Marja L. Anderson, "Modern China: South," in *Food in Chinese Culture: Anthropological and Historical Perspectives*, ed. K. C. Chang (New Haven, CT: Yale University Press, 1977), pp. 295–316, 316–82.

10

Born from Bears and Corn

Why Indigenous Knowledge Systems and Beliefs Matter in the Debate on GM Foods

Shiri Pasternak, Lorenzo Mazgul, and Nancy J. Turner

Introduction

Indigenous peoples do not have a common religion, but almost all share a history of colonialism and Christian missionization. In spite of legal and political colonial impositions, cultural practices relating to food production and consumption have been central to preserving and transmitting to future generations local ecological knowledge, social institutions, ethnic identity, and spiritual teachings. Food practices, including prohibitions, evince an extraordinary body of knowledge passed down over generations orally and through observation and participation. Today these practices have been increasingly threatened by a widespread transition from locally produced and prepared food to the consumption of marketed, globally sourced, refined, and processed foods (see Letourneau's chapter). According to Debra Harry, director of the Indigenous Peoples Council on Biocolonialism, genetically engineered foods represent to indigenous peoples worldwide both the extension of an ongoing colonial destruction of their local knowledge systems and a violation or desecration of the natural world.[1]

A succession of declarations by indigenous peoples has identified appropriation of genetic resources and indigenous intellectual property as threats to their permanent access to and sovereignty over natural resources. The Declaration of Indigenous Organizations of the Western Hemisphere (Phoenix, Arizona, February 1995) asserted, "[w]e oppose the patenting of all natural genetic materials. We hold that life cannot be bought, owned, sold, discovered, patented, even in its smallest form." The Ukupseni Declaration (Kuna Yala, Panama, November 1997) stated, "[w]e reject the use of existing mechanisms in the legalization of intellectual property and patent systems . . . including intellectual property rights and patents to legalize the appropriation of knowledge and genetic material, whatever their source, and especially that which comes from our communities." Finally, the International Cancun Declaration on Indigenous Peoples (5th WTO Ministerial Conference, Cancún, Mexico, September 2003) demanded that the polity "stop patenting of life forms and other intellectual property rights over biological resources and indigenous knowledge."[2]

The relationship between colonialism and the biotechnology revolution was expressed, for example, in the declaration of the United Nations Development Programme Consultations on the Protection and Conservation of Indigenous Knowledge in Sabah, Malaysia, in 1995, which stated, "The indigenous people's struggle for self-determination is a very strong counterforce to the intellectual property rights system vis-à-vis indigenous knowledge, wisdom, and culture. Therefore, the struggle for self-determination cannot be separated from the campaign against intellectual property rights systems, particularly their applications on life forms and indigenous knowledge."[3]

Vandana Shiva, an adamant critic of the GM "revolution," also sees biotechnology as an extension of colonialism, a process that trades largely on the ostensible supremacy of Western forms of knowledge. Shiva writes, "[t]he duty to incorporate savages into Christianity has been replaced by the duty to incorporate local and national economies into the global marketplace and to incorporate non-Western systems of knowledge into the reductionism of commercialized Western science and technology."[4] GM research from this perspective is an example and enlargement of this forced incorporation: "The colonies have now been extended to the interior spaces, the 'genetic codes' of life-forms from microbes and plants to animals, including humans."[5] In other words, "[t]he creation of piracy through the piracy of others' wealth remains the same as 500 years ago."[6]

What Crosby has called the "Columbian Exchange"[7]—the global swapping of plant germplasm between the New World and the Old[8]—ultimately meant that every single species of economic significance in America has benefited from introgression of foreign genes. According to Jack Kloppenburg, over a period of about four hundred years, the world has seen a "global and unprecedentedly rapid movement of plant germplasm, a process that has been shaped in important ways by an ascendant capitalism committed to the creation of new social forms of agricultural production worldwide."[9] And while germplasm from the South is considered "the common heritage of humankind," the proprietary technology of genetically engineered seeds bears a significant price tag. What entitles corporations to patent "codes of life" or the genetic sequences of plants and animals is the elevation and privileging of labor that generates profit, protected by intellectual property law, over thousands of years of collective, public improvement undertaken by indigenous farmers around the world.

Since most of the justification for GM food development rests on the promise of "feeding the world"—meaning the developing world, where the greater percentage of GM foods are *not* grown—and most "food insecure" people live in the Global South and therefore have been subject to colonial rule and could be identified as indigenous, these claims must be weighed against indigenous knowledge and belief systems around food. Therefore the discourses of development and hegemonies of scientific truth that prefigure and justify technological solutions to what critics argue are social, spiritual, and political inquiries must be called into question here.

The first task of this chapter, however, is to adequately define the millions of people in the world regarded as "indigenous" and tentatively connected through the vagaries of history. The United Nations has defined "indigenous peoples" as "communities, peoples and nations . . . having a historical continuity with pre-invasion and pre-colonial societies that developed on their territories, who consider themselves distinct from other sectors of societies now prevailing in those territories, or parts of them."[10] It is difficult not to conceive of indigenous peoples in relation to colonial European conquerors or modern attempts at economic, social, and political restructuring and assimilation. However, it is not helpful to think of indigenous peoples as opposite to an industrial, modern society, either. "Indigenous" connotes a *dynamic* people who are ancestrally, spiritually, and politically connected to a territory in a multiplicity of ways.

Although there exist differences among nations, and between indigenous peoples within nations or tribes, a common position among indigenous people globally is often indicated where matters of ecological health are concerned. For example, a questionnaire was developed by the International Indian Treaty Council (IITC) to elicit responses from indigenous peoples around the world focused in part on the relationship between traditional cultural practices and food systems. There were 128 respondents from 28 countries or states around the world. Findings showed that despite massive heterogeneity among participants, a common position upheld by indigenous peoples was the essential connection between traditional subsistence foods and practices and the maintenance of their respective community's culture. Biotechnology (including genetically modified organisms, or GMOs) was cited as a current practice that results in negative impacts to their communities' traditional foods and food practices.[11]

Indigenous food practices are tied inextricably to sustaining ecological health, often through food taboos, prohibitions, and spiritual teachings that impart a respect for the integrity of the earth and its life forms. These complex forms of knowledge and socialities represent a worldview that integrates and informs community organization based on people's direct dependence on the natural world. These forms of knowledge are often referred to as Traditional Ecological Knowledge and Wisdom (TEKW). In contrast to conventional scientific paradigms, TEKW represents "[p]ractices of aboriginal people to maintain and enhance their lands, waters, and living resources derived from generations of experimentation and observation, leading to an understanding of complex ecosystems and physical principles."[12] The earth is not considered in this paradigm to be merely the property of those currently tending to its guardianship, but rather, for example as Mohawks believe, that it is to be managed with respect for future generations (seven generations), as a common heritage.[13] Two central aspects of this ecological management and social practice are kincentricity or "animistic" worldviews,[14] and following that, a belief in and deep respect for the spiritual interconnectedness of all things.[15]

Although the concept is somewhat contentious in that there is no common understanding of what TEKW means and how it should be implemented in policy and public arenas, what seems clear is that TEKW forms a special category for research. Unlike conventional science, this knowledge is understood "not as something separate from its possessors' lives, but as something integral to the individual."[16]

The authors selected two research methods we felt most effectively narrated *particular* indigenous perspectives on transgenes in food, rather than treating indigenous peoples as a unified entity. Lorenzo Mazgul and Shiri Pasternak traveled to Guatemala to conduct focus groups in Mazgul's hometown of Patzún with the local Maya community. Nancy Turner sent out questionnaires to indigenous colleagues in North America. The questionnaires, as well as the Guatemalan focus groups, were designed to elicit responses regarding the role of traditional foods and food gathering practices in indigenous cultures. More specific questions were then asked about perspectives on the role of transgenes and GMOs in relation to traditional food-gathering practices and the belief systems in which they are embedded.

Case Studies

There is little doubt that the current demand for resources by the global human population has gone beyond the earth's carrying capacity. The depletion of resources and the need to produce more food for a growing population are key contributors to this demand. The pressure on the world's resources is compromising the food security of many people, especially those in developing countries. The typical scientific and positivist approach to address the consequences of the scarcity of resources, however well meaning, is that key measures such as investments, scientific research, transfer of knowledge, and technology—presumably from developed to developing countries—will "ensure wise management of resources and sustained capacity for growth."[17]

In contrast to the scientific and positivist approach, and despite the incredibly rapid expansion of Western economic and ideological approaches to resource use, some indigenous peoples worldwide continue to utilize natural resources that achieve the same goals of "wise management of resources" and the ability to feed themselves, yet the means to achieve these goals are not drawn from scientific approaches, but include rituals, ceremonies, and prohibitions.[18]

The ongoing challenge for indigenous societies, because of the expansion of Western economic, cultural, and ideological perspectives, is whether they can maintain their own traditional belief systems and practices, not just for the sake of preserving them but also as a means of ensuring their own physical survival and the continuation of social and time-tested ecological relationships.

The case studies for this chapter focus on the town of Patzún, Guatemala, and a cross-section of indigenous peoples, mostly scholars, from North America whose work is related in some ways to traditional food practices. Considering that corn (*Zea mays*), the most important food for the Maya, is one of the major crops that the biotechnology industry has invested most in,[19] it is appropriate and important to find out from the people themselves their views on genetically modified foods. Their perspectives will help others to understand whether they accept consumption of GM corn and other crops and whether measures such as labeling are sufficient to guarantee their right to know about and have control over the foods they eat. Corn is also a central foodstuff for many other indigenous peoples of the Americas. Considering that Canada and the United States are major centers of food biotechnology in the world, corn and other traditional foods may be susceptible to cross-contamination with GM seeds. Furthermore, in their day-to-day diets, indigenous people may also consume a good deal of processed food containing GMOs.

Until recently, most farmers in Patzún have grown corn intercropped with beans (*Phaseolus vulgaris*) and squash (*Cucurbita pepo*), using low-input traditional methods. In the 1980s, crops such as broccoli (*Brassica oleracea*) and snow peas (*Pisum sativum*) were introduced as an economic development measure; today, these crops are cultivated with high inputs of synthetic pesticides and fertilizers. These characteristics of Patzún, where the modern Western approach to agriculture converges with the largely traditional approach used by the local Maya population,[20] constitute an important rationale for focusing on this community and an opportunity to understand how the local people react to, think about, and accept or reject genetically modified foods.

Indigenous peoples in North America, owing in part to epidemics and colonial massacres that resulted in a massive population decline estimated to have killed around 90 percent of their peoples, as well as to a systematic cultural genocide that involved Christian residential schools and government appropriation of their traditional lands, face challenges of a more protracted colonization and assimilation. However, we are witnessing today the restoration and revitalization of North American indigenous cultures and the transmission of their knowledge of traditional food practices to the younger generations. In this light, GM foods represent a paradigmatic case for the conflict between indigenous and nonindigenous worldviews. Those North American indigenous peoples we asked to respond to our questionnaire were known to us as people who have a specific interest in food. Thirteen people participated, including six males and seven females. Of these, eight different North American indigenous nations were represented, including Cherokee Nation, Wasco (Warm Springs Confederacy and other nations), Gwich'in, Métis, Mohawk, Haida, Nuu-Chah-Nulth (three different communities) and Straits Salish (two communities).

There is also significant research with indigenous peoples worldwide on the issue of GM foods. For example, the Maori of New Zealand, Zambian farmers in Africa, and Philippino *campesinos* all have strong movements of opposition to the technology, in an effort to maintain sovereignty over their lands and to protect the nature upon which they depend to subsist.[21] We have undertaken these focus group discussions and interviews to examine more closely such positions.

The Role of Traditional Food in Indigenous Diets: Sowing, Cultivating, and Harvesting

The town of Patzún is a typical community in the rural highlands of Guatemala, with a large Maya population that has retained a distinct identity and speaks the Maya language Kaqchikel. The majority of the Patzún Mayas are Christians; therefore, at times when the participants refer to "god," it is not clear whether they are referring to the Christian god, one of their own contemporary conceptualizations, or a god of pre-Hispanic origins. Despite these ambiguities, the importance of corn for the survival of Mayas weaves through and interlocks their history from their origins up to today: "According to our culture, the food that sustains us is corn. This is our basic but necessary food. Often when we talk among ourselves, we comment that our stomachs are not made for other types of foods like bread [from wheat—*Triticum aestivum*]. We were raised eating corn, which our ancestors left us, and because corn originated from here, we can practically say that *we were born from corn*" (emphasis ours).[22]

Corn is recognized as sacred, not just in its consumption, but also in the rituals, ceremonies, and blessings that accompany its sowing and harvest. Corn illustrates the historically indivisible meanings of corn as a food, a spiritual worldview, a form of ecological management, and a community practice: "Our ancestors have taught us that before you plant the corn, you have to go [to] the field to give thanks and ask permission from the land. They taught us that the land has a god, its divine owner, so planting corn was not just a simple and free act, you have to give thanks and ask permission from the God. This is what I keep in my mind from the teaching of our grandmothers and grandfathers."

Descriptions of corn rituals also underscore a kincentric worldview. Corn is described here not as an object, but in familial, animate relation as "our mother." Reciprocal obligations are thus involved:

> Another thing that I have learned is that the day before the harvesting of corn takes place, the harvesters visit the field to be harvested; they take candles, burn them, and call on the corn so that they are in the fields on the day of the harvest. People say that our mother, the corn, may go wandering and leave the fields and may not be there on the day of the harvest. By visiting the cornfield the day before harvesting and asking our mother to be in the field on the day of the harvest, you make sure that all the corn is harvested and nothing is left on the fields. Also, some people . . . have a big meal once the harvest is finished—bring a live chicken to the field and slaughter it right there, cook it and then have a big feast; all this to give thanks to the land for the favor done for this year of providing food.

Until twenty to thirty years ago, the Mayas subsisted on their traditional diet. Changes in this diet mean a loss of ecological knowledge and rituals that teach respect for what the earth, their mother, has given them: "These rituals are being forgotten because some people do not have time to practice them. The lesson of this belief is that all the corn must be harvested and nothing is to be left on the field."

A critique of colonialism also runs throughout the responses. The introduction of new lifeways and new foods has brought many problems for the Mayas: "The *g'eka* [white people and mestizos] have tricked us into eating foods that they eat. We are now copying them, but these foods give us illnesses; we now have illnesses that in the past we did not have. We think that the white man's food is better, that it tastes better, but all it has done is give us illnesses." Dietary changes contradict the tried gastronomical knowledge of the Maya, as well: "We do not eat like the g'eka, who eat several types of food one after the other, we are not like them. We have been taught, and passed from generation to generation, that there are good foods, and for us the main food is the blessed corn, this is what sustains us." The notion of plenty, valued by consumers in the market society in the form of product diversity and "eating strawberries all year round," holds little stock with an elder interviewed in Patzún: "If we only had tortillas to eat, we feel we have enough food, even if we only have salt to eat with our tortillas. This is how we view things. For us, foods such as corn and beans are sacred, and we considered them as sacred seeds, which have been left for us by our ancestors."

All of the North American indigenous questionnaire respondents use traditional food (food that was eaten by past generations of their people) to some extent, and all of them indicated a preference for traditional food. Jaalen Edenshaw (Haida) emphasized why this was important to him: "It is important to continue gathering and eating traditional food because it is from the land, and we are from the land. It is important to know where your food is coming from, to kill it, handle it, and prepare it. You get a respect for where it comes from and what it provides to you when you are a part of it." Those living away from their peoples' territories reported that they eat their traditional foods less often, though, because their foods are not available. This "food distancing" marks here a substantive break from cultural practices.

Environmental problems, industrialization, and settlement expansion due to population growth have affected peoples' ability to access traditional food, as well. As explained by Pakki Chipps of the Victoria, BC, Canada area: "I would like to use traditional foods entirely, but we have a serious shortage of them as the aggressive introduced plants erase the traditional food plants. We have traditional places where our food plants grew and where we would travel in order to gather them, but we no longer have access to these places, or they are just paved or damaged too."

All of the respondents indicated that food and dietary practices[23] were "important" or "extremely important" to them. They provided a range of specific details about cultural traditions relating to food gathering and use, from ceremonial practices around harvesting, to ways of caring for the food-producing areas. Alestine Andre, a Gwich'in woman from the Northwest Territories, Canada, explained how cultural and spiritual practices, as well as ecological management techniques, are embedded in this knowledge of food-gathering practices: "Respect must be used when harvesting or gathering traditional foods. Cultural practices must be followed on the land. For example, when collecting plant foods, the gatherer or harvester must give a token of thanks (wooden match sticks, tobacco, tea leaves, etc.) before [he or she] pick[s] any parts of the plants, including their berries. Care must be given to ensure there are food resources for the future."

Earl Claxton Jr., of the Saanich Nation, Vancouver Island, described the importance of salmon fishing for him, bringing to light the political significance of traditional food practices: "I think for me it is very important to go and catch the spawning salmon in the Goldstream River. This allows me to be visible to the general public while I am practicing my treaty right."

Moreover, the respondents constantly emphasized the social aspects of food gathering, preparation, and eating: "When you are preparing or cooking country foods, share some cooked food with an elder, widow or widower, or another person" (Alestine Andre). Another respondent illustrated the link between cultural transmission and food practices: "Like most families we eat together daily; however, when it is fishing season, we generally spend a lot of time with extended family, fishing, smoking, and jarring fish, and eating fish with our grandparents, grand aunts and uncles, etc. . . . This is also a time when the younger ones are taught to do fish" (Tah-eh-soom-ca). Changing lifestyles have affected the ways in which people interact, though. One respondent described how since many people are no longer food gatherers because they are busy making money now, families spend more time apart.

In Patzún, the importance in food practices of spending time with others was also highlighted, as was the erosion of this tradition and the social forms of life that accompany the practice: "In our culture to eat with the family is very important; it is a special time. . . . The family is a strong unit; eating together is very important because at this time we all share with each other how our day was and tell each other what sort of things we have been doing."

Foods for Human Consumption That Are Avoided Because They Are Considered Unacceptable

There were very interesting similarities between the nature of food taboos expressed by the Mayas in Patzún and by the indigenous questionnaire respondents in North America. For example, pregnant women were subject to many food prohibitions. In Patzún, "Pregnant women should avoid eating citrus fruits such as lemons [*Citrus X limon*], oranges [*Citrus sinensis*], pineapple [*Ananas comosus*], nance [*Byrsonima crassifolia*], and avocado [*Persea americana*]. These foods can cause abortion, and some can result in allergies. For example, if a pregnant woman eats avocados her child could become susceptible to colds, running nose, etc. So pregnant women have to be very aware of what they eat; they have to look after themselves."

Another Maya respondent spoke of food taboos for pregnant women in terms of animal consumption: "Something that we have been told about is that we should not feed chicken legs to a boy because this makes him not have the ability to climb a tree. Also, some people believe that if a boy eats chicken wings or if a woman who is pregnant with a boy eats chicken wings, when the boy grows up he can turn out to be violent towards his wife; this is because people believe that chicken wings can cause you to not think clearly, you can lose your head."

One respondent in the North American group, a young adult, noted that there were traditional food taboos, such as those around avoiding certain foods during pregnancy, but that her generation did not necessarily observe them.

Seasonal prohibitions were also a common feature and at times were also connected to pregnancy or lactation. Pakki Chipps noted that there were plants that could be used only during certain times of the year, for very specific purposes. She explained, "[s]ome of the plants prohibited to young women, pregnant women, and lactating women have contraceptive qualities that could harm the fetus. Some foods, if picked at another time than that specified as correct, would die out if picked at any other time."

The theme of kincentricity also arose in the questionnaires. Jeff Corntassel (Cherokee Nation) and Taiaiake Alfred (Mohawk) both noted that people should not eat their own clan animal, and in particular, Jeff Corntassel explained from the perspective of the Cherokee Nation, "[b]ears are our brothers. These people started as part of a now nonexistent Bear Clan. They prayed to ask the Creator how they could best serve the people, and the next morning, they were turned into bears. Eating a bear is like eating a fellow *Tsalagi*."

A point that cannot be emphasized enough here is that what is taken for granted in science as the *natural order* is called into question by indigenous worldviews on a deep epistemological level. Modern taxonomic systems reflect the historical contingency of scientific truths and the hegemony of these categories when they delineate animal, human, and plant kingdoms. For example, Jeff Corntassel, in responding to the question about food taboos, stated that "foods from the upper world (i.e., birds) should not be mixed with food from the underworld (i.e., fish). These boundaries should also be respected." This point also pertains to Corntassel's associations between creatures and "forces of nature" in the world: "Rattlesnakes come from where Thunder (*U-hv-yv-da-gwa-los-ti*) resides in the West and should never be disrespected or eaten."

While some may be tempted to call these taboos and prohibitions "superstitious"—such as the Maya belief that young girls should avoid pineapple, orange, avocado, and lemons as they can delay puberty and interfere with menstruation (thus unfounded and vestigial), this accusation is itself recognized by scholars as an increasingly outdated position. Not only have many of these traditional beliefs proven to be scientifically rigorous and significant discoveries,[24] but the destruction of local knowledge systems around the world has often been done in the name of an advanced, superior science that has caused more damage than good. For example, the synthetic pesticide DDT was first introduced as a miracle product for increasing agricultural yield. Decades later, it was discovered that DDT was accumulating in toxic levels in the water table, the soil, and the bodies of humans and animals and that it would take years, maybe centuries, for this damage to be repaired.

Pakki Chipps reinforces this point: "Some foods [taboos] may be just superstition, but as much of this knowledge comes from very wise ancestors, I am not about to discount it as superstition just because I don't always know the reason for the taboo. The knowledge of plants and their qualities is extremely ancient and was built upon for perhaps thousands of generations." She added, "I trust the ancestors far more than I trust the grocery store or the advertisers or the WTO."

The gesture here is to recognize science as a story *among others*, not as a hegemonic truth. This idea was eloquently stated in the Maya study group: "What our ancestors believed in and have taught us is useful for us to survive; when we forget these beliefs, it affects our ability to survive." An important lesson here is that the colonial imposition of hegemonic truth has endangered these crucial knowledge systems: "Being exposed to religion [Christianity], which some of us may not understand well, has affected some of our beliefs. Sadly, beliefs that have been forgotten could be very useful for us now; but even if we wanted to recover them, it may not be possible because they have been abandoned for too long."

Thoughts about Genetically Modified Foods and Genetic Material from Unacceptable Foods Being Introduced into Otherwise Acceptable Foods

Common themes emerged in our surveys that captured the difficult ethical questions that GM foods pose to some of the most marginalized people in the world today. Articulated often as a set of oppositions, the desire to improve or perfect nature was weighed against the potential of unintended negative consequences; the choice to accept GM foods was weighed against its current imposition; and the violation of cultural beliefs was weighed against the relentless demand of economic imperative.

In the North American respondent group, the majority (eight) reject-ed outright the notion of eating GM food under any circumstances. The introduction of DNA or genetic material from prohibited foods to oth-erwise acceptable foods through genetic engineering techniques was com-pletely offensive to participants. From a political perspective, Taiaiake Alfred responded succinctly: "I do not like the idea of genetically modifying plants or animals to make them more marketable or to increase profits for corpo-rations. I don't think they are poisonous or harmful, but I think that they disturb the natural balance and promote the overgrowth of capitalism and corporate power." Jeff Corntassel stated, "It would be unacceptable and a violation of *Ani-geel-ahgi* natural law," and Jules Chartrand explained, "I say leave the Creator's work alone. I trust his work better than man's alterations of his work." Pakki Chipps was most emphatic, bringing to bear, as many of the respondents did, indignation toward what was perceived as yet another project imposed upon indigenous peoples without consent, interlaced with concerns for the protection of nature and indigenous culture: "I would feel violated. But we have been raped culturally, linguistically, physically, men-tally, so why not genetically too? I already feel violated knowing that the store-bought foods I am forced to buy because our traditional foods are not available are altered; for them to introduce genetic materials from plants or animals that are taboo would be yet another attempt at cultural genocide. Certainly, it would be the ultimate insult and sign of disrespect for our communities and traditions."

Another point to consider in the debate is the way the kincentric worldview held by many indigenous peoples situates animals and plants on the same level as humans so that to genetically modify them would be akin to genetically modifying ourselves. As Pakki Chipps argues, "We should not have the right to tamper with plants and animals, for they are people too!"

Jeff Corntassel explained his position for distinguishing the modifica-tion of sacred foods as a distinct issue: "If I know [that a food has been genetically modified], then I can make a better determination. It ultimately depends on what the food is. If it's a sacred plant like corn, for example, I won't tolerate any genetic modifications. If it's a domesticated animal, like a pig, I might tolerate some small modifications. Overall, I would tend to avoid these genetically modified foods and view having our own seed banks and hunting wild animals to be a form of resistance to the arrogance of *yonega* (white) genetic prospectors. *Tsalagi* principles for leading a good life urge our people to eat our foods from well-tended gardens. How can we call our gardens 'well-tended' if they are based on modifications of our sacred plants and derivatives of sacred animals?"

He also explained that "there are deeper cultural reasons for not eat-ing these animals [*Tsalagi* foods] other than a slight modification in their appearance or livelihood. Modification of this type would only demonstrate disrespect for that animal or plant's role in giving life and further undermine the balance of natural law. *Tsalagis* have an origin story of *Selu* (Corn Mother) and *Kanati* (The Hunter). . . . It is this balance of roles and responsibilities that keeps our communities healthy and vibrant. To interfere with these roles via genetic modifications is to threaten the well-being of our communities and future generations."

In Patzún, the question of transgenes in acceptable foods was also articulated in terms of the relationship between health concerns and cultural beliefs. For example, one respondent stated, "I think that if we know that the foods have been genetically modified we would not eat them, especially those foods that affect our health and cultural beliefs. For example, if a pregnant woman knows that a food has genes from a rabbit or a dove she would not eat them."

In Patzún, the idea that GM foods can *improve* on nature is not always antithetical to God or the Creator's will; however, these changes must be approached in terms of principle: "I think it is important to know about them [genetically modified foods] because I think that according to the scientists, the genetic modification of food is to improve the plant and animal species for our consumption. However, we are all conscious that we have our principles, some of which are very spiritual, and we know that God created things that are in themselves perfect. We have to stop and think about these new things carefully. In recent years, there have been new things that have been beneficial to us, but others have been harmful and have had negative consequences."

However, if one is given no choice over the consumption of GM foods, God or the Creator's will cannot be taken into account: "I will add that we always try to choose the foods we consume. We consume tomatoes [*Lycopersicon esculentum*], we consume fish, but we choose to eat them. If these two things are mixed together, then we do not have the choice. I think that genetically modified foods can be good economically for those who cultivate them; they can benefit the producer because these crops could yield more. But we have our principles, and we think that God has created everything in nature, and we should respect what has been created."

Sacred foods, which can vary from nation to nation, were often cited as the foods for which genetic modification would cause the greatest offense. They may be "perfect" foods, or gifts from the Creator, and they may also economically and socially sustain a group for centuries. The reasons for such sacred status can confound a "rationalist" worldview. As Umeek (Nuu-Chah-Nulth) elegantly states, "[t]raditional foods are not limited to logic and mathematical precision but can be considered sacred and life bringing. The salmon is one such food. The salmon gives its life in exchange for recognition and honor, attributes not necessarily amenable to logic."

Such sacred foods and the natural processes that they represent and embody, though, are often pushed aside when economic imperative meets cultural beliefs. In Patzún, one community member stated,

> One of our main concerns with genetically modified foods is that they are the result of human intervention, that they are not the result of natural processes. To us the natural process is sacred, and to interfere with it is not acceptable, so we think that genetically modified foods in general would not be acceptable for consumption. However, we also have to be realistic, especially in communities like ours, where many of us are poor and can only afford to buy certain things. Therefore, if in the future the only foods available here are all genetically modified foods, we would not have much choice but to consume them, even if we did not accept them, because we would not be able to go and look for other foods elsewhere, and probably more expensive. We do, however, feel that we should know whether a food is genetically modified or not.

Economic imperative was also cited in the North American group where two respondents indicated that they might consider eating GM food under some situations (for example, in a case of financial necessity). "However, I believe that in the end economic necessities will be the final determinant for consuming genetically modified foods. People starving will probably eat them, especially when our population numbers will be so high and food demand will be high." The broader economic context of GM foods was brought up when Jaalen Edenshaw expressed, "I've heard the argument that it is the only way to feed the poorer people of the world, but if GM foods are owned by private industry they will be trying to make a profit; where is the profit in the poor nations?"

However, it must be stated that despite reservations expressed about the science of GM foods, some indigenous peoples were open to considering beneficial aspects of the technology. Umeek (Nuu-Chah-Nulth) considered some potential benefits of GM products. He identified potential benefits to crops: enhanced taste and quality; reduced maturation time; increased nutrients, yields, and stress tolerance; improved resistance to disease, pests, and herbicides; and new products and growing techniques. For animals, he acknowledged that genetic modification could provide the following: increased resistance, productivity, hardiness, and feed efficiency; better yields of meat, eggs, and milk; and improved animal health and diagnostic methods. Some environmental benefits of GM products he identified included "friendly" bioherbicides and bioinsecticides; better conservation of soil, water, and energy; bioprocessing for forestry products; and better natural waste management. And finally, for society at large he noted that GM products could increase food security for growing populations. On the other hand, Umeek noted a number of concerns with GM products ranging from their potentially negative impacts on human and environmental health and on access to food, to the ethical problems they raise regarding intellectual property, and to social justice internationally. Umeek summarized his views as follows: "Genetic modification, as with any process that involves transformation, appears to be a two-edged sword. One edge appears to be beneficial and the other edge dangerous and risky. Considering the incipient and risky nature of GM foods I think it unacceptable that the U.S. . . . is not required to label GM foods, and consequently there may be no way, at present, to gauge the long-term effects on health and well-being of populations that are unwittingly consuming such foods."

In Patzún, farmers reflected, "Given what you have said about genetically modified foods, to me, at first it appears that they may be good. For example, if we can grow tomatoes that are resistant to frost we could benefit a great deal if we produce a lot of tomatoes and we are able sell them; however, I am concerned about their consumption. What would the consequences be? Would there be any health impacts? Would it result in illnesses?"

Labeling

In the North American focus group, all twelve respondents were emphatic that all GM food must be labeled so that people can inform themselves and make their own choices about whether or not to consume these foods. One respondent noted that "the spirit of the food is changed as well as the nutritional content."

In Patzún, the focus group participants were also unanimous in their desire for labeling: "I think that if genetically modified foods are introduced here, they should be labeled and have a warning about their possible consequences. If you know that a food is genetically modified, it is your own decision to consume it or not, but if you do not know, then you do not have a choice, and you cannot make your own decision. Maybe in places where these foods are being created, people there do not even consume them; maybe they are fed to animals." The unknown long-term health risks of GM foods were also cited as a key factor in the desire to see GM foods labeled:

> If we consume these foods, maybe a few years down the road we
> will suffer negative consequences. It is possible that we will not
> see consequences two to three years from the time we consume

them, but over a long time, the consequences will appear and damage a person's health. Even now, I think we are suffering the consequences of modern things. Before our time, our grandmothers and grandfathers had a long life expectancy. The average life expectancy a long time ago was one hundred years, then it became eighty years, then sixty; it has slowly been decreasing, and I think this is because the foods in our diet are no longer pure; they have been tampered with, and we use chemicals and interfere with God's creation.

However, due to the pressures of population, these people may not have a choice, whether or not there are labels: "I will add that our situation today is such that we have enough to feed ourselves, but given the rate at which our population is growing, in the near future we will lack food to feed ourselves, and at that time we will need to find other ways to feed ourselves. Instead of starving, we may have to consume genetically modified foods. I think that at times of great need no matter how much warning we are given about the danger of consuming something we will still consume it."

Conclusion

Both the responses to the questionnaire and the discussion from the focus groups affirm that most indigenous peoples, because of their unique histories and interactions with local ecosystems, have distinct cultures, languages, religions, and social and economic organizations. Their food systems are likewise unique, and many indigenous peoples today strongly identify with particular culturally important foods, including methods of production and preparation and customs observed in serving and consuming them. More than most members of society at large, indigenous people are embedded in the ecosystems that produce their food and often have a better appreciation of and direct experience with their food sources, whether harvested from the wild or domesticated and cultivated.

Indigenous peoples' recognition of their embeddedness in and interdependence with the ecosystem leads most of them to hold a strong belief that components of the ecosystems are sacred, that is, that time and space and all living things that exist within them are things with which harmony must be maintained, that they are not to be tampered with. This belief in the sacredness of most things provides direction for the way they conduct their lives as well as shapes their world. For many, the violation of the natural law through interference in natural process is unacceptable; therefore, the genetic modification of foods is unacceptable.

Through their beliefs and practices, indigenous peoples stress that their dependence on the ecosystem is both physical and spiritual. Ceremonies and rituals are conducted at specific times and places to present offerings to the life forces of fertility, rain, plants, and animals; they also are ways of communicating with the spirits of these life forces. For the Mayas of Patzún, corn is the mother, the lifegiver; communicating with her before harvesting is vital for a good crop. Do genetically modified foods have a spirit with which communication can occur? For some indigenous people, human interference with the genetics of foods makes them "genetically mutilated," as one respondent described it.

The holding of ceremonies and rituals at specific times and places assures effective communication with the spirits. The cornfield and other places where food is gathered or hunted become the religious altar; the times of sowing and the times before, during, and after harvesting become the religious altar. Will the cultivation, gathering, or raising of genetically modified foods be conducive to ceremonies and rituals at specific times and places?

The types of foods eaten, or not eaten (taboos), during ceremonies, rituals, and everyday life are observed for various reasons: to recognize the interdependence with the ecosystem, to respect and honor the wisdom of ancestors, to communicate and give thanks to the spirits of life forces, and to ensure human survival. In the case of the Mayas of Patzún, many of the food prohibitions concern the assurance of a good crop, a check on population growth, and care for the unborn and future generations. Not knowing what genetically modified foods contain clearly affects the Mayas' belief of what foods should or should not be eaten.

There is a general feeling from the respondents to the questionnaire and the participants in the focus groups that genetically modified foods are unacceptable, not just for consumption but also because of how they affect the sacredness of the natural order. However, recognizing the socioeconomic forces that propel the propagation of genetically modified foods, they demand the mandatory labeling of these foods so that at least people have the choice of whether to consume them or not.

Acknowledgments

We wish to thank the following indigenous participants for their careful consideration of our questions regarding their use of GM food: Christine Aday (Nuu-chah-nulth Nations, Ucluelet and Hesquiaht); Dr. Taiaiake Alfred (Kanien'kehaka Kahnawakero:non—Mohawks of Kahnawake); Alestine Andre (Gwichya Gwich'in, Gwich'in Nation); Caroline Chartrand (Red River Métis Nation); Jules Chartrand (Red River Métis Nation); Pakki

Chipps (Coast Salish); Earl Claxton Jr. (Saanich Nation, Tsawout); Dr. Jeff Corntassel (Cherokee Nation); Tah-eh-soom-ca (Dawn M. Smith) (Nuu-chah-nulth Nations, Ehattesaht); Umeek, (Dr. E. Richard Atleo) (Nuu-chah-nulth Nations, Ahousaht); and an anonymous member of the Confederated Tribes of Warm Springs, Oregon, with other cultural affiliations.

In addition, we thank the men and women of Patzún who participated in the focus groups for sharing their wonderful stories, strong beliefs, and insightful comments.

Notes

1. "Biopiracy and Globalization: Indigenous Peoples Face a New Wave of Colonialism," *Splice* 7, no. 2 (January/April 2001), p. 3.

2. Declaration on Indigenous Peoples' Rights to Genetic Resources and Indigenous Knowledge, 6th Session of the United Nations Permanent Forum on Indigenous Issues, May 14–25, 2007, New York, New York.

3. Cited in Rosemary J. Coombe, "Intellectual Property Rights, Human Rights & Sovereignty," *Indiana Journal of Global Legal Studies* (fall 1998), p. 10.

4. Vandana Shiva, *Biopiracy: The Plunder of Nature and Knowledge* (Boston: South End Press, 1996), p. 2.

5. Ibid., p. 3.

6. Ibid., p. 2.

7. Alfred Crosby, *The Columbian Exchange* (Westport, CT: Greenwood, 1972).

8. As Kloppenburg elaborates: "The New World supplied new plants of enormous culinary, medicinal, and industrial significance: cocoa, quinine, tobacco, sisal, rubber. More than this, the Americas also provided a new arena for the production of the Old World's plant commodities (e.g., spices, bananas, tea, coffee, sugar, indigo)." *First the Seed* (Cambridge: Cambridge University Press, 2004), p. 154.

9. Kloppenburg, *First the Seed*.

10. Sustainable Development Networking Programme (SDNP), "International Day of the World's Indigenous People." http://www.sdnpbd.org/sdi/international_days/Indigenous-people/2004/index.html.

11. Final Report on an Indigenous Peoples' Initiative to Establish Cultural Indicators for SARD: Questionnaire on Indigenous Peoples' Traditional Foods and Cultures. Distributed by the International Treaty Council (IITC) and submitted to the United Nations Food and Agriculture (FAO) Rural Development Division (SDAR), (August 25, 2003).

12. Nancy J. Turner, Marianne B. Ignace, and Ronald Ignace, "Traditional Ecological Knowledge and Wisdom of Aboriginal Peoples in British Columbia," *Ecological Applications* 10, no. 5 (2000), p. 1276.

13. Ibid. Also, see Haudenosaunee Kanienkehaka Mohawk Nation Roots of Citizenship, http://www.kahonwes.com/blood/citizen.htm; http://www.kahonwes.com/language/kanienkehaka.html.

14. See Eugene S. Hunn, "The Value of Subsistence for the Future of the World," in *Ethnography: Situated Knowledge/Located Lives*, Virginia Nazarea, ed. (Tuscon: University of Arizona Press, 1999), pp. 23-36; also E. Salmón, "Kincentric Ecology: Indigenous Perceptions of the Human–Nature Relationship, *Ecological Applications* 10:5 (2000), pp. 1327–32.

15. Nancy J. Turner and E. Richard Atleo (Chief Umeek), "Pacific North American First Peoples and the Environment," in Harold Coward, ed., *Traditional and Modern Approaches to the Environment on the Pacific Rim: Tensions and Values* (Albany: State University of New York Press, 1998), pp. 105–24.

16. George W. Wenzel, "Traditional Ecological Knowledge and Inuit: Reflections on TEK Research and Ethics," *Arctic* 52, no. 2 (June 1999), p. 117.

17. Rosamond L. Naylor, Walter P. Falcon, Robert M. Goodman, Molly M. Jahn, Theresa Sengooba, Hailu Tefera, and Rebecca J. Nelson, "Biotechnology in the Developing World: A Case for Increased Investments in Orphan Crops," *Food Policy* 29 (2004), p. 16.

18. Nancy J. Turner and Fikret Berkes. "Coming to Understanding: Developing Conservation through Incremental Learning," *Human Ecology* 34, no. 4 (2006), pp. 475–78. See also Stephen J. Lansing, *Priests and Programmers* (Princeton, NJ: Princeton University Press, 1991).

19. Cf. Naylor, et al., "Biotechnology in the Developing World."

20. Lorenzo Mazgul, "The Environmental Sustainability of Non-traditional Cash Crops in the Highlands of Guatemala: A Focus on a Maya-Kaqchikel Community," masters thesis, University of Victoria, 2004.

21. See the film produced by the Indigenous Peoples Council on Biocolonialism, *The Leach and the Earthworm* (2003).

22. This and other quotes in this section are from two focus groups conducted in Patzún, Guatemala, in May 2005.

23. The questions, more specifically, were about how the food is produced, how and where the food is bought, where the traditional foods are gathered or harvested, how food is prepared, and how, when, and with whom food is eaten.

24. Fikret Berkes, *Sacred Ecology: Traditional Ecological Knowledge and Resource Management* (Philadelphia: Taylor and Francis, 1999).

11

Regulatory and Innovation Implications of Religious and Ethical Sensitivities concerning GM Food

Conrad G. Brunk, Nola M. Ries, Leslie C. Rodgers

The analyses of the various religious and ethical "food cultures" offered by the authors of the chapters in this book graphically illustrate how critically important food is in the life of a people. The cultures included in this study were chosen because they have particularly strong and specific ethical and religious concerns about the kind of food they eat and about the way that food is produced, prepared, and consumed. Many cultures and religions we have not examined in this study would have similar sensitivities.

For most people and in most cultures food is far more than simply nutrition and sustenance. This is true even when food is in scarce supply and people face the constant challenge of malnutrition and even starvation. In such circumstances people may be forced by necessity to violate their most strongly held convictions and sensitivities about what is appropriate to eat, but they do not thereby abandon those convictions and sensibilities. This attitude emerged occasionally among focus group participants who eschewed the notion of genetically modified foods but acknowledged that practical exigencies might force them to relinquish their ideals—with regret:

> If I were to go hungry and my children . . . and only GM prod-
> ucts were available . . . I'd eat them. But if it's a choice, I say
> no. (Jewish participant)

[W]e also have to be realistic, especially in communities like ours, where many of us are poor and can only afford to buy certain things. Therefore, if in the future the only foods available are genetically modified foods, we would not have much choice but to consume them, even if we did not accept them, because we would not be able to go and look for other foods elsewhere." (Maya participant)

The act of eating is one of the two most intimate activities in which human beings (and other animals) engage. It is the act of absorbing into one's body elements of the external world, and in the process one exposes oneself, not only to the pleasures and benefits inherent in that exposure, but also to a host of risks, diseases, toxins, and allergens, to say nothing of various aesthetic shocks.[1] Throughout human history this intimate exposure of one's body to the external world, like sexual intimacy, has been imbued with profound symbolic, and therefore religious and ethical, significance. And, like sexuality, it has been carefully controlled and regulated, not only or even primarily as a means of ensuring survival in the face of great risk, but also as a means of defining cultural and religious identity. In most traditional cultures food carries a profound symbolic significance; how and what one eats expresses fundamentally who one is and the fundamental goods to which one is committed. A culture's conception of the good life, or of the just and righteous life, is reflected in its rules and habits about the production, preparation, and consumption of food—in what is or is not "good for you."

Often this culinary conception of the good is expressed in terms of "naturalness." So, what many focus group participants considered to be good for you was food that is natural—as created and provided by God or Mother Earth. In fact, some viewed "tinkering" with the fundamental makeup of food to be a form of technological hubris.

[W]e know that God created things that are in themselves perfect. We have to stop and think about these new things carefully. (Maya participant)

I object when mankind says we can improve on the things that are already here. . . . [I]t's definitely arrogant. (Jewish participant)

I think we need to have a certain degree of humbleness and respect for the way that things have been created. And that falls away when we start to play Creator ourselves. (Mennonite Christian participant)

In religious cultures particular foods themselves can take on powerful symbolic meaning, raising them to the status of the sacred. Many of the authors in this volume have pointed this out in their accounts. In Hinduism, for example, rice takes on this status, as does congee (also made from rice) in the traditional Chinese culture. In the Christian tradition, bread and wine carry profound sacramental significance. In many aboriginal cultures, most food carries a sense of the sacred attachment to the land and the animals from which it has been taken. For many indigenous Mexicans, maize represents a sacred ancestral heritage. Foods such as maize in Mexico, rice in China, or bread in Europe, which are the "staff of life" in the culture, typically carry this kind of significance. Technological or other nontraditional impacts on highly symbolic foods are often seen as forms of "contamination" or "impurity," understood religiously and morally.

It might be expected that these symbolic meanings would have lost their power in modern, technological, and secular societies, and this may be largely true. One of the oft-cited impacts of modern technology is its tendency to desacralize the world by visualizing everything in it in instrumental, utilitarian terms.[2] This tendency is probably most advanced in the fast food culture of North America, where food is viewed by many in instrumental terms, as fuel to be ingested as quickly and efficiently as possible to sustain the other activities of life that really count. In the fast food culture, food becomes homogeneously bland in character, and eating loses much of the ritual significance involving family and fellowship with friends that it holds in most traditional societies. It becomes a mere commodity, divorced from the social networks of its production on the farm, its processing, and even, in the case of ready-to-eat foods, its preparation.

Regardless of this tendency, it may be the case that food is one of the last elements of secular, technological societies to be recognized for symbolic significance. Observers of European and North American food safety regulation have pointed to examples where industry or regulators ignored or discounted the symbolic power of certain foods much to their peril. It is widely acknowledged that the public outcry in the United States several decades ago when the USDA approved the use of the chemical pesticide spray ALAR, a suspected carcinogen, stemmed from the fact that its primary pathway to human exposure was as a residue on apples. Failure to recognize the symbolic significance of the apple as "health food" in American culture ("An apple a day keeps the doctor away") led regulators to treat ALAR as an acceptable risk. The public concern in North America about the use of the synthetic BST hormone in cows to stimulate greater milk production is widely attributed to the symbolic power of milk in North American culture as an essential source of life and health, particularly of children. Canadian regulators, recognizing this symbolic power, found a way to ban the use of BST in Canada.[3]

The concerns initially raised about genetically modified foods in many countries around the world should not have come as a surprise to anyone who understands the powerful symbolic significance of food and diet. It is not by accident that this opposition is strongest in societies that retain the strongest cultural attachments to food and weakest in those that have largely lost these attachments. Many people within biotechnology industries themselves have recognized the mistake of having chosen food crops as the first generation of genetically modified plants, and especially to have chosen to engineer into them characteristics such as pesticide and pest resistance that benefit primarily producers rather than consumers.

DNA, Darwin, and Food Purity

The chapters in this book examining the religious, ethical, and symbolic significance of food within different religious and moral communities raise important questions about the interests and attitudes affecting consumer acceptance of new food technologies, particularly those foods that are the product of genetic engineering. They also raise questions about the regulation of these food products by governments, particularly questions concerning the kind of information available to those for whom the hidden characteristics of the food they purchase and eat are critical to their acceptance of them on moral or religious grounds.

These chapters focus mostly on issues related to the transfer of genetic material, or DNA, from one organism to another by the use of genetic engineering techniques. New biotechnologies permit the transfer of genes across species and kingdom barriers that do not or cannot normally occur in nature apart from these technological interventions. Because they do not normally occur in nature, they are not part of the normal system of expectations within which food-related beliefs and practices have evolved within cultures.[4] Thus, it should not be surprising if these transfers raise serious questions about the contamination of foods considered acceptable for human consumption by elements of plants or animals that are not acceptable. In a culture or religion that considers a particular plant or animal to be morally objectionable or obnoxious as food, the mixture of elements of that organism into acceptable food is bound to raise questions. The fact that the element in this case is a portion of what the scientists themselves commonly call the "blueprint" of an organism, which is thus understood by laypeople to give the objectionable organism its character (its DNA) does not alleviate these questions but in fact exacerbates them. This should not be particularly surprising.

The chapters of this book illustrate that when it comes to the food they eat, people tend to draw very clear lines between the acceptable and the unacceptable, even though they may find it difficult to explain why they draw the lines they do. The reasons for these distinctions can be explicitly religious, ethical, or purely aesthetic. Often they are all involved, as was evident in the responses of the focus group participants.

> Is it really ethical for people to mix these two things? [animal and vegetable genes] . . . [O]bviously there will be more rabbinic discussion about this, but I think for me it's more of a visceral reaction of taking the characteristics of the fish and passing it on to something that isn't fish. . . . [T]here is no reasonable reason for reacting to that. (Jewish participant)

> [I]t's . . . offensive [to transfer a fish gene to a tomato]. . . . [I]n essence the fish has been violated—the integrity of that creature . . . there's a sort of . . . denigration. (Vegan/vegetarian participant)

As Laurie Zoloth points out in chapter 4, traditional Jewish dietary rules may seem to the outsider, and even to many internal rabbinical scholars, to have an element of arbitrariness in them: what is the "essential" difference, after all, between a fish with scales and one without scales or a mammal that chews its cud and one that does not? Why is one acceptable and the other not? But to the devotee the distinctions are salient and critical.

Such distinctions are especially puzzling to the modern scientist or philosopher, whose understanding of the world is shaped by Darwinian conceptions of species evolution that blur the borders between species, making them open and constantly changing. This understanding is bolstered by the discovery of DNA and RNA, the chemical structures whose combinations orchestrate the differences and similarities among all organisms. In this understanding, the idea that certain base pair combinations are in any way essential to one species or type of organism, to say nothing of "inappropriate" if placed in another, makes little sense. Neither does it make sense in this conception of things to say that a particular sequence on the double helix of DNA that contributes code for a particular characteristic in an organism—that is, a gene—is essential to that organism. If this gene is taken from a lobster (a prohibited food in the Jewish and Muslim religions) and spliced into the genome of a sunflower, it does not make sense to call it a "lobster gene" in any sense other than as a reference to the means by which it was obtained. There is nothing lobster about the particular gene itself that would be transferred to the sunflower. So from this perspective it makes no sense to suggest that the newly modified sunflower has lobster or even lobster product in it. DNA is simply DNA, wherever it is found.

One of the interesting paradoxes reflected in the chapters of this book is the sometimes stark difference between the attitudes of the lay members of the religious and moral communities we studied and those of the scholars who reflected more philosophically upon the place of genetically modified foods in their community's theological or moral understanding of food and diet. The scholars, not surprisingly, reflect much more strongly the Darwinian, postessentialist ontology of DNA representative of the contemporary intellectual world, especially of the West. For them, the idea that a genetically engineered food product might be rendered unacceptable by a transgene from some "unacceptable" animal or plant was less compelling than it was for most of the lay members of the focus groups. For the latter, even among those with a fairly sophisticated understanding of the science of genetic engineering, the presence of DNA taken from unacceptable food sources had much greater significance. There was a strong sense that something of the lobster itself would be transferred to the sunflower. Though it was rarely articulated explicitly by persons in the focus groups, the underlying assumption seemed to be that if the gene in the lobster contained part of the information that made it a lobster rather than something else (i.e., it was information *unique* to that organism), then transferring the information coded for in the gene to another organism involved a transfer of something lobster into the host organism.

At work in these different responses to the issue is a deeply philosophical, or ontological, question. It could be characterized as a primarily materialist, nonessentialist ontological orientation versus an idealist, essentialist one. The Darwinian, modernist view is the former, reflected in the impulse that "DNA is DNA" (wherever it is found and whatever proteins it codes for). It is not essentially different whether in the lobster or the sunflower. The viewpoint of the laypersons in the focus groups seems to run much more in an idealist, essentialist direction, placing much greater significance upon the reality of what is coded for in the DNA. In this view "DNA is not just DNA." What is coded for in the sequences of DNA is seen as what makes all the difference in living things, and it is what is ultimately the "reality" of things—of the lobster on the one hand and the sunflower on the other. This is why to many participants in the focus groups it seemed obvious that the presence of a lobster transgene (for example) in their sunflower seeds (for example) meant that there was something of a lobster in their sunflower seeds. If, within their religious or ethical perspective, eating lobster was a problematic activity, then it was perfectly conceptually consistent to wonder whether there might be an issue with a lobster transgene in their sunflower seeds.

Within a more scientific perspective it often looks as if the unwillingness of a Jew, Muslim, or vegan to eat sunflower seeds with a lobster transgene or tomatoes with a pig transgene is based on an unscientific assumption that the sunflower has become a lobster or the tomato a pig. But, there is no reason to attribute such a scientifically or philosophically naive assumption to these concerns. Although the focus group participants were not asked to comment specifically on that issue, it is clear from the discourse that they did not consider the food with the transgene from a prohibited food animal or plant source to have actually *become* that prohibited animal or plant. There was rather more of a sense that the presence of the transgene constituted a form of *contamination* in the otherwise acceptable animal or plant. The notion of "purity" appears to underlie many of the attitudes toward acceptable and unacceptable food in religious and moral dietary frameworks,[5] and the possibility of transgenes from impure plant or animal sources engenders this sense of contamination. This sentiment was a key theme running through almost all the focus groups.

> Moderator: So if there was a gene from an animal in a tomato, that would then not be considered a pure food.
>
> Response: Not considered pure—
>
> Moderator: Contaminated?
>
> Response: Exactly. . . . [W]e'd consider it adulterated or contaminated, foreign, something foreign. (Exchange with Hindu participant)
>
> Moderator: What's the threshold for you? How many genes? Let's use the example of the tomato . . . how many genes from a pig in a tomato plant would make that tomato unacceptable for you?
>
> Response: One.
>
> Moderator: And what would concern you about it?
>
> Response: Well, the fact that my belief as a Seventh-day Adventist is that pork or pig is unclean, that tomato is now unclean because it has that gene in it. (Exchange with Seventh-day Adventist participant)

There is nothing particularly arcane or mysterious about such concerns with the purity of food. This concern for food purity is increasingly
evident, even in largely secular societies and subcultures. It is part of the
impulse driving increasing demands in these societies for organic food on the
grounds that it is uncontaminated by pesticide residues, antibiotics, growth
hormones, or chemical additives. Even when scientists point out that scientific evidence does not suggest any serious health risks associated with these
substances, many people still maintain their preference for organic products
on the basis of their greater purity or naturalness. Scientific evidence of risk
is not the sole criterion for choice of food and it is not just religious folk
who are concerned about purity.

Taking Food Sensitivities Seriously

The accounts of the cultural and religious sensitivities around food quality
and food purity provided in the other chapters of this book show clearly
why concerns about technological alterations of food or food production
techniques go much deeper than mere passing personal preferences, whims,
or tastes. To treat them as mere whims or preferences, as they often are
in the typical economic analyses underlying marketing and public policy
decisions, is to seriously underestimate their significance and their potential
power in the market.

Food sensitivities are commonly rooted in deeply held religious and
moral philosophies and embedded in long traditions, and they represent
fundamental life commitments, moral principles, or aspects of self-identity.
Failure to take these into account in the development of new food technologies and the transfer of these technologies to the marketplace can lead to
serious disruptions and inefficiencies. The opposition that the introduction
of the first generation of genetically modified foods met in Europe, Africa,
and parts of Asia is an example of this kind of disruption. This opposition
was anticipated neither by the industries that developed them nor by the
governments who approved them for the market. As consumer demand for
labeling of such products reverberated through the international markets,
major producers in North and South America suddenly found large markets
virtually closed to them because of the economic costs of segregating and
labeling GM and non-GM food products.

The so-called first-generation GM food products were simple single-
gene transfers (the transfer of one single gene from one organism to another
to confer a single desired trait) that did not alter the quality, negatively or
positively, in any way apparent to the consumer. The traits conferred to the
plants involved were largely those beneficial to the producers of the GM

crops, conferring pest resistance[6] or resistance to general herbicides,[7] both of which were said to decrease the costs and environmental risks of pest control. The fact that the biotechnology industry chose as first-generation products those that had no discernible benefit for consumers is cited by many observers as a major strategic error in obtaining consumer acceptance. If these products had improved flavor, shelf-life, appearance, or even cost, then, it was argued, consumers would have been less likely to have seen themselves as exposed to health or environmental risks from which others (e.g., the corporations and farmers) derived the primary benefits. Had they been the beneficiaries of the technology, they would have been far more accepting.

There may be some truth to this analysis. But it would be naïve to believe consumer preferences and behavior are this simple or can be manipulated so easily. The analyses provided in the chapters of this book should put to rest any facile assumption that promises of benefits of taste, cost, or other aspects of food quality are sufficient to determine consumer responses to new food technologies, especially for those consumers who bring to the market profound religious, moral, and cultural understandings of food and diet. One of the clearest messages to emerge from the focus groups engaged in the research for this book was that the first generation of genetically modified food products, as controversial as they were, did not begin to engage the kind of concerns associated with the dietary restrictions of major religious and cultural communities. This is because these products did not involve gene transfers of the type that would trigger the specific moral and religious sensitivities around food associated with these communities. From this point of view they were relatively benign, involving plant-to-plant or bacteria-to-plant transfers that raise few issues beyond those of the technology of genetic engineering itself.

However, in principle, opposition to genetically engineered food was not of greatest concern in the focus groups. Nor did the chapter authors who reflected philosophically upon food practices in these traditions find within them any basis for such an "in principle" opposition. Instead, both the lay members of the focus groups and the scholars of the traditions were more concerned about certain kinds of genetic manipulations and transfers. Particularly worrisome, especially among the lay persons, was the transfer of animal genes into plant foods or the genetic modification of animal food products themselves. This is not to say that the transfer of plant genes, especially from plants not considered appropriate as food within certain traditions (e.g., Hinduism) did not raise similar issues. But the introduction of animal genes into plant foods, or of genes from "forbidden" food animals into other animals or plants, was far more troubling for people. This reaction was amplified for those with specific food taboos based on their religion or ethics.

[A]ny part of the pig or pork . . . is unlawful. And even sometimes when the food that we eat touches bacon or pork or
ham . . . we wouldn't eat it. . . . So of course when it's genetically modified . . . even a very small amount of it, it still makes
it unlawful [to eat]. (Muslim participant)

I think as a vegan it's totally unacceptable to have an animal's
product to modify something that I would eat, totally unacceptable. (Vegan/vegetarian participant)

Rejection of plant-to-plant genetic transfer, on the other hand, was
based far less on food taboos than on concern about the consequences of
interfering with nature.

Moderator: Let's say a gene from a broccoli plant into a carrot
plant . . . Is that still violating natural order?

Response: In a sense it is. Alteration is—non-interference . . . it's
a very big, huge word in our belief . . . we believe there's a
natural order . . .

Moderator: So if humans interfere and do something they're
putting that harmony—

Response: In a twist. (Taoist participant)

Thus, as future generations of genetically modified foods are developed,
and the types of gene transfers become more complex, involving transgenes
from more varied plant and animal sources, "gene stacking" (the transfer of
multiple genes to produce multiple traits), and more visible changes in plant,
and especially animal, physiology, it is likely that consumers will have much
greater concern about the nature of the food available to them in the marketplace. Consumers, especially those with concerns rooted in the religious,
moral, and cultural traditions explored in this book, will demand more information about the genetic character of the foods they purchase. Without such
information, they could be expected to lose confidence in the food products
available to them in the dominant markets and create demands for "niche"
markets that can assure them the food they purchase is acceptable within
the terms of their religious or ethical convictions and practices.

The emergence of the strong niche markets for organic foods as well
as natural health products in opposition to the dominant food and therapeu-

tics markets are examples of this dynamic already at work. In increasingly multicultural societies like those in Canada, the United States, and Western Europe, the numbers of people who reflect the kinds of dietary concerns documented in this book will also increase, and the demands for the wide availability of "pure" foods (however each culture or religion defines pure) will likely intensify. The growing sensitivities about the use of animals for food, and the resulting growth in the ranks of persons who adopt a principled ethical vegetarian, or vegan lifestyle is hugely significant in itself.

The vigor with which some groups will assert their right to choose pure foods was highlighted by litigation in the United States in 1998 that challenged the presumption expressed in the Food and Drug Administration's (FDA) policies on GM foods that foods created through recombinant DNA technologies are generally recognized as safe and do not require special regulation.[8] The plaintiffs, who included representatives of Christian, Jewish, Buddhist, and Hindu faiths, as well as public interest groups, argued, among other things, that the lack of a regulatory requirement for labeling such foods violates religious freedoms protected by law in the United States. Relevant U.S. federal law does confer some power on the FDA to require labeling and defines food labels as deficient if they fail to reveal facts about the material "with respect to consequences which may result from" using the product. The plaintiffs argued that "materiality" should not just be restricted to safety concerns, but should extend to other consumer interests, including religious views related to acceptable foods. However, the litigation eventually failed as the court ruled that the existing legal doctrine mandates deference to the FDA's policy choices, and the lack of labeling did not pose an undue burden on the exercise of religious beliefs. Although the case was unsuccessful on largely technical legal grounds, it nonetheless demonstrates the strength of religious convictions related to food and the willingness of groups to defend their religious interests through legal action.

Consumer sensitivities regarding food based on religious, moral, and cultural beliefs pose a set of serious questions for the future of food technology in society. The assumption that the biotechnologies and nanotechnologies of the future are mere extensions of the food technologies of the past, and will continue to be widely accepted, is not likely to be reliable. For many of the people who practice the dietary strictures analyzed in this book, technologies that blur the boundaries between plant and animal and between species of plant or species of animal will pose serious problems. The new technologies will very surely blur those boundaries. Unless consumers are provided with reliable tools for making choices consistent with their food-related values, their only authentic option is to opt out of the food markets dominated by these new food technologies.

The GM Food Labeling Debate

There has been a vigorous debate in many countries about whether food products containing genetically modified plant or animal material should be labeled as such. The policies governing GM food labeling have differed widely in different countries and regions. Citizens of the European Union (EU) countries have successfully lobbied for some of the most rigorous labeling requirements for GM food products. In the EU it is mandatory that all foods containing genetically engineered plant or animal products be labeled as such.[9] Other countries, such as Japan and China, have similar labeling rules. India has recently passed legislation requiring the mandatory labeling of all foods containing ingredients derived from biotechnology or bioengineering, or food products containing GM material. Further, no GM foods may be sold in India unless a Genetic Engineering Approval Council has approved them.[10] In North America, GM foods have been introduced into the marketplace with little public knowledge or debate. Although public opinion polls are consistent in their findings over the past decade that a strong majority of respondents (typically over 80 percent)[11] favor mandatory labeling of GM foods, governments have not felt obliged to adopt mandatory labeling regulations. In the United States and Canada, the vigorous opposition of the food and biotechnology industries to the labeling of GM food products has won the day at the level of public policy.

The members of the focus groups in this project strongly mirrored what the opinion polls in most developed countries have consistently found among the general populations of those countries. While not necessarily opposing all use of genetically modified organisms in food, they strongly believed that as consumers, and particularly as members of religious and moral communities, they deserved at least to have essential information about the nature of the food products they purchase easily available to them. Indeed, focus group participants passionately believed that consumers have a right to know what is in the food they consume and expressed the need for various levels of information. At a minimum, they called for clear labeling of food products containing genetically modified constituents or the equivalent, labeling denoting "GMO-free" products. Many people with particular food prohibitions also wanted to know the source of the transgene.

> I don't eat pork. . . . I'm not going to buy food that has that in it, and I should be . . . informed of exactly what I am putting in my body. (Seventh-day Adventist)

> I think that it has to be specific . . . where [the transgenes are] coming from and what kind of modification is done. (Jewish participant)

For most of these people it seemed obvious that the most effective and efficient means of providing this information was to require food producers to state clearly on labels not only whether a food contained a genetically modified plant or animal product but, if so, from what source the transgenes in the product were obtained. Clearly, if a significant number of people for whom the genetic character of their food is vitally important can participate fully in a marketplace only if they have access to information about the origins of transgenes in that food, there must be compelling reasons for not making such information available.

Industry and government opposition to mandatory labeling of GM food products has been based on a number of considerations and arguments. Among the most common are the following:

1. The long-standing philosophy of food labeling in North American regulatory contexts has been that food labels should be used to provide only two kinds of information: (a) information about the nutritional content of the product; and (b) information relevant to the health and safety of the product. In the latter case, for example, foods containing ingredients that can cause allergic reactions, such as various types of nuts or chemical preservatives, should be labeled to indicate the presence of these potential allergens. Wines and other products carry warnings about the risks of alcohol consumption to pregnant women. Since genetically modified products are required to be tested for human safety before market approval, and since the process by which the foods were grown or processed is irrelevant to their nutritional content, there is no reason for the label to indicate the process by which they were produced. An indication on the label that the product has been genetically modified adds no information relevant to either nutritional content or food safety.

2. Because people have come to expect that a label will carry only nutritional or safety-related information, the argument is frequently made that including the information on the food label that the product contains genetically modified plant or animal ingredients will communicate to consumers the erroneous message that there are safety or nutritional problems associated with genetic modification techniques. This would lead to unfounded fears among consumers about GM foods that might significantly reduce their sales.

3. Requiring the labeling of GM food products would greatly increase the costs of food. It would require the complete

segregation of GM plant and animal products through the entire production chain, from "the farm to the fork." GM corn and soybeans, for example, would need to be kept separate from conventional corn and soybeans on the farm (including protection from cross-pollination between them), transported in dedicated trucks and railroad cars, stored in separate bins, processed in separate facilities, and so on. The maintenance of essentially two separate food production systems would greatly increase the costs of production, and these costs would be passed down to consumers. It is not fair to those who desire the benefits of GM foods, nor is it economically efficient, so the argument runs, to add these costs to consumers because of the desires of a minority of consumers to have this information.

4. There are complex problems involved in finding acceptable criteria for what counts as a "GM" or a "GM-free" food product. Since most commercial food products are processed foods containing a host of ingredients, there will be many cases in which the major ingredient may be a non-GM plant or animal, but one of the minor ingredients (e.g., a preservative or a flavoring) will be GM. Should this be considered a genetically modified food? How much GM content, in other words, should be permitted in a product that is labeled as non-GM? Or what about the case in which the product is made from a genetically modified plant or animal, but itself contains no trace of DNA (or of the specific transgene) and is chemically indistinguishable from its non-GM counterpart? For example, oil extracted from GM canola is chemically indistinguishable from the oil from non-GM canola (as long as it does not contain traces of the canola plant, which will carry the DNA). If this oil is one of the ingredients in a packaged food product, should it be labeled as GM or not? Opponents of GM labeling cite these complications as politically too difficult to settle to the satisfaction of all consumers.

The proponents of mandatory GM food labeling claim there are serious flaws in each of these arguments. With respect to (1), they point out that, in fact, there are many precedents in North American policies of labeling requirements that are related neither to health nor to nutrition. One of these is the requirement in the United States that the place of origin, or source, of a food product be on the label so that consumers can decide whether to buy foods from other countries or to support the domestic American industry. This seems a clear case of mandatory information on a label designed to enhance consumer power to exercise social and moral judgment in the marketplace (i.e., whether to support domestic or foreign products). In the European context, it has been pointed out that the EU adopted mandatory labeling for GM food explicitly to enhance consumer choice.[12]

The U.S. requirement that irradiated foods be labeled as such is another oft-cited example of a food production process that does not affect the final character of the product. It certainly does not, according to the scientific consensus, add any significant risk to health. Indeed, food is subjected to irradiation to decrease the microbiological risks to consumers. The requirement for the labeling of irradiated foods is the result of strong public demand, based on associations with the health risks of direct human exposure to radiation (which does not in fact occur with irradiated foods).

Proponents of mandatory GM food labeling reject argument (2), above, as well. They point out that any unsubstantiated fears about the safety of GM foods based on the label are purely a result of the continuing claim of regulators that labeling should deal only with health and nutritional matters. Further, they point out that even if these labels cause concerns about the safety of GM products, do consumers not have the right to make their own choices about the safety of GM products? If they have concerns about these products, why should they be prevented from exercising their concerns in the marketplace just because the producers and regulators of the products consider their fears to be unfounded? Interestingly, EU regulations for labeling of GM foods have provisions that require disclosure on the label of any characteristics of a food product "where a food may give rise to ethical or religious concerns."[13] This sets a precedent for including label details to provide information for those who hold religious or ethical views that shape their food choices.

Few people contest the validity of the assumption in argument (3) that mandatory labeling is costly and that the increased costs would surely be passed on to consumers. However, it may be a cost the proponents of mandatory labeling are willing to pay. They point out that these costs would be spread equitably among all consumers, which they should be. The alternative, for those who do not want to eat certain GM foods, is to shop in highly specialized markets (e.g., certified organic, kosher, halal, etc.), in which all the added costs of segregating these foods and guaranteeing their non-GM status are passed on to this small subset of consumers. These consumers are likely to view the introduction of GM crops, animals, and food products into the food chain with no discrimination, segregation, or labeling as the imposition of a costly burden upon them. The GM industry reaps the benefits of unregulated integration of their product into the market, while the costs are "externalities" passed on to those who do not want them. This is a major issue, covered in the discussion later on voluntary labeling.

The counter to argument (4) against mandatory labeling—that it is too complex and controversial to implement in nonarbitrary ways—is that reality disproves the claim. It is in fact possible to arrive at standards for the allowable minimum content of transgenic material in food products and to have reliable means of testing for this content. It is possible also to arrive at agreements on the question of whether products derived from GM plants or animals that do not contain transgenic DNA and are chemically indistinguishable should be labeled GM or not. The reality that disproves the claim is the fact that the EU and other jurisdictions have reached political agreement on these issues and have implemented such standards.[14] Further, these same problems need to be solved if standards for the voluntary labeling of GM foods are adopted, as they have been in Canada.[15]

Voluntary Labeling

Voluntary labeling has been proposed as one option to address consumer demands for information about GM foods in the marketplace, and jurisdictions such as Canada and the United States have adopted voluntary labeling rules. In these countries, regulators do not distinguish between GM and non-GM foods in imposing labeling requirements; rather, mandatory labels are required only where the nutritional composition of a food product is significantly changed or where heightened risk for toxicity and/or allergenicity exists. However, recognizing that some companies may wish to make claims regarding the GM status of food products, voluntary labeling guidance has been developed to describe acceptable claims that are truthful, not misleading, and verifiable.

In April 2004, the Canadian General Standards Board, a governmental agency, promulgated the *Standards for Voluntary Labeling and Advertising of Foods That Are and Are Not Products of Genetic Engineering*.[16] This voluntary standard was developed with participation of government, consumer, and industry groups and aims to provide guidance for food manufacturers and importers on acceptable label claims for GM and non-GM foods. In the United States, the Department of Agriculture issued a draft voluntary labeling guidance document in 2001, *Guidance for Industry: Voluntary Labeling Indicating Whether Foods Have or Have Not Been Developed Using Bioengineering*.[17] Some states have initiated legislation to mandate labeling of GM foods[18] and at the national level, the GE Right to Know Act was introduced in Congress in 1999.

Several factors may impede the adoption of voluntary labels. First, manufacturers of foods that are a product of genetic engineering have little incentive to label the product as such because some consumers may perceive the label as a "warning" and choose not to purchase the product. As mentioned earlier, consumer research reveals a consistent trend in the majority of survey respondents indicating they are less likely to purchase GM-labeled products (but other research suggests these stated preferences are not borne out in actual purchasing decisions).[19] Further, in countries with mandatory GM labeling laws, GM products have reportedly disappeared from store shelves, arguably diminishing overall consumer choice. In some European countries, food retailers themselves adopted voluntary bans on GM ingredients before mandatory labeling laws came into effect.[20]

Consumers who wish to avoid any genetically engineered ingredients may be dissatisfied with the allowance in labeling rules for accidental (adventitious) inclusion of ingredients from GM crops. The current threshold in Canada and Japan for adventitious content is up to 5 percent, and the EU has adopted a 1 percent threshold, which has been criticized as requiring unattainable standards for segregation of GM and non-GM crops. Manufacturers who wish to label their products as not being products of genetic engineering may have difficulty meeting some label standards. For example, a manufacturer may label corn chips as "made entirely from non-genetically engineered ingredients" only if the origin of all ingredients can be verified. Due to concerns about inadvertent contamination of crops, it may be difficult to verify that ingredients are absolutely free of GE ingredients. Finally, a voluntary labeling regime can be subject to criticism if there is no system for monitoring compliance.

Monitoring for compliance was an important issue emerging from the focus groups. Overall, participants viewed the motivations and trustworthiness of the GM food industry with a fair degree of skepticism, and a number of people said they would not count on the government to monitor industry properly:

> It's money-driven completely. . . . It's for corporations to make money, and . . . they are not telling people the whole truth. (Jewish participant)

> Moderator: So there are some issues of trust around government.

> (Several voices) Oh yeah. Yes.

Only a fool would trust the government. (Vegan/vegetarian participant)

Internet Bar Code Sites

To provide consumers with additional information regarding food products, some advocate bar code systems that would permit shoppers to scan products to look up websites with additional information about the characteristics of GM products (similar to scanning devices that are currently in use to allow shoppers to check prices of products).

Bar code systems could enhance traceability and consumer confidence and choice. Precedents already exist for providing enhanced access to information about food sources for consumers; for instance, information technology systems in some Japanese supermarkets allow shoppers to look up information about meat products to determine their origin:

> Walk into a Jusco supermarket in Yamato, a small city near Tokyo, Japan, and you can glimpse the future of meat. In addition to a conventional bar code, each steak package sports its own ID number. Type the number into the computer sitting on a nearby table, and up pops information about the cow the steak came from: a scanned copy of its negative test results for mad-cow disease and, in case you are interested, its breed and sex, its date of slaughter, and the name of the producer. At some Japanese meat counter displays, you'll even see a picture of the family that raised the animal.[21]

While this technology has been developed to allay consumer concerns around disease risks—particularly in regard to beef and "mad cow" disease—it could conceivably be extended to allow consumers to look up information about whether food items are products of genetic engineering and to look up risk assessment information.

At present, some jurisdictions have developed websites that provide searchable databases of regulatory reviews of various agricultural biotechnology products. For example, the U.S. Department of Agriculture, the FDA and the Environmental Protection Agency have established the US Regulatory Agencies Unified Biotechnology Website[22] that includes regulatory reviews for crop plants. A search for "cantaloupe" reveals two entries for bioengineered plants that have been modified to achieve delayed fruit-ripening traits. The website provides links to FDA documentation summarizing the nutritional and safety assessment of the fruit. In the European Union, the Community Register of GM Food and Feed[23] provides information on GM foods that have been authorized under EU law for entry into the market, including links to risk assessment documentation. Technologically savvy focus group participants identified online information systems as useful resources. "[A]nother thing that would be very helpful even though we may not understand it is . . . what gene has been transferred, so that if we want to look it up on the internet, we can find out the properties or the traits of the genes that have been transferred into that product" (Seventh-day Adventist).

It may be argued that implementing in-store bar code scanners and maintaining electronic databases on products would be costly, may serve to overwhelm rather than inform consumers, and may "stigmatize" GM foods. However, technological requirements for such information systems already exist, as do publicly accessible government databases containing information on approved GM food products, so the cost implications may not be prohibitive. Indeed, stores that invest in such technology may enhance their market share as information-seeking consumers change their purchasing patterns to shop at stores that offer access to information regarding a food's origin and characteristics. Consumers who are not interested in this level of detail are free to bypass it and proceed directly to the checkout counter. Such technology need not be limited to GM products; in fact, significant investment in "farm-to-fork" tracing technology is aimed at providing consumers with information about the safety and quality of meat.

Consumer Choice and Respect for Religious Practice

The debate about the mandatory labeling of genetically modified foods is usually seen as an argument between economists and utilitarians on the one side, who see the problem primarily as that of trying to make markets as productive and efficient as possible, and consumer advocates on the other side, who see it more as an issue of respect for consumer autonomy or freedom of choice. According to the utilitarian view, rational individuals seek to maximize personal satisfaction through choice by selecting the course of action that has the best chance of producing an outcome consistent with their personal preferences. The best system is the one that optimizes these preferences, whatever they are, and whatever their basis. If some individuals prefer non-GM products, a food system in which this option is available will better serve consumer preferences than one in which this choice is unavailable.[24] There is no question that focus group participants, no matter what their religious, cultural, or ethical persuasion, felt that respect for consumer choice is a fundamental right:

[W]e should at least be given the opportunity to know . . . whether we want to buy this food or not buy this food. And that's what bothers me. (Seventh-day Adventist)

[T]hey should label things. . . . We should have a consent [choice] too. (Theravada Buddhist participant; this group was generally open to GMO)

Moderator: How would you feel if you knew that a food that you had eaten . . . was genetically modified and contained a gene that had come from an animal?

I would feel pretty upset about that. Not due to the fact that I've consumed it but more about not knowing it. I would feel . . . violated. (Vegan/vegetarian participant)

I think that if genetically modified foods are introduced here, they should be labeled. . . . If you know that a food is genetically modified, it is your own decision to consume it or not, but if you do not know, then you do not have a choice, and you cannot make your own decision." (Maya participant)

I think the more people know about what they're consuming, the better choices that they make. And I think it is unfair to withhold that information. (Mennonite Christian participant)

But the utilitarian analysis also puts this preference on an equal footing with other consumer preferences, such as the desire for inexpensive or long-lasting foods. Indeed, many economists would argue that a food system that did not allow those who wanted to purchase less expensive GM foods to act on this preference would be as problematic as one that denies the choice of GMO-free, if not more so. Further, the confusion produced by a complex system of labels and consumer information would likely substantially reduce consumers' overall ability to satisfy their preferences. Thus, the utilitarian approach to the issue of choice requires a complex weighing of the costs and benefits that would be associated with labeling. This weighing generally comes out against the added costs and complexities of mandatory labeling.

Consumer advocates generally make the argument for mandatory labeling on the basis of ethical arguments that emphasize individual autonomy and rights of choice. The argument here is that the greater the freedom of choice in the marketplace the better, regardless of the rationality of the choices people make. People have their own reasons for deciding which products they wish to consume or not consume. This argument for consumer autonomy is often bolstered by appeals to capitalist and free market ideology: a market is not truly "free" unless consumers have the information they need to exercise their values and preferences. Here, the underlying claim is that markets should not be constructed so that consumers are unable to act in accord with their basic values and beliefs.[25]

The argument from autonomy carries greatest weight when it is asserted on the basis, not just of the right to act in accord with one's preferences, but *the right to act in accord with fundamental values and commitments.* When food choices and practices represent values that are of deep importance to individuals—importance rising to the level of religious and moral conscience or identity, such as those practices articulated in the various chapters of this book, the autonomy argument can be bolstered by the appeal to rights of freedom of religion and conscience. These latter rights are recognized as deserving special protection, not only in liberal political theory, but also in the legal frameworks of nearly all modern liberal states. A system of choice in the marketplace that constrains a person's ability to act on the basis of fundamental religious and moral values compromises a society's commitment, not only to respect for autonomy generally but also to respect for freedom of conscience.

Skeptics regularly point out that individuals often deviate in their consumer behavior from their espoused religious or ethical convictions about food and diet. This can easily be interpreted as evidence that these preferences are much weaker than the individuals or their religious advocates claim. However, respect for autonomy must also include the freedom to follow or deviate from values fundamental to personal and cultural identity. It is one thing for individuals to violate such beliefs freely and something entirely different for society to develop a system of practices that forces them to do so.[26]

Conclusions: Addressing Regulatory Implications of Religious, Cultural, and Ethical Food Practices

The chapters of this book make a strong case on the basis of respect for religious and moral conscience for some market structure that respects consumers' rights to have the information they seek about the genetic characteristics of the food available to them in the market. The accounts of how people from these religious and ethical traditions think about food, food purity, and the appropriate means of food production, as well as how they think about the genetic makeup of the food they eat clearly establish that these attitudes and practices should carry great weight in the structuring of food production and marketing.

The following implications can be reasonably drawn from the analyses of the religious and ethical food cultures in these chapters:

1. The religious and ethical traditions examined in this book represent the major religious traditions of the world. Hundreds of millions of the world's population consider themselves to be devotees and practitioners of these religions. This represents a significant influence in the marketplace.

2. Nearly all of these religions have well-defined rules or expectations placing limitations on the production, preparation, or consumption of food, and these are at the very heart of the practice of the religion or ethical culture itself. It is not easy to identify oneself as a devotee without observing these food practices. These practices will surely manifest themselves in consumer acceptance of new food technologies.

3. DNA is important: for many practitioners of the food and dietary traditions represented in the major religious and ethical vegetarian/vegan cultures of the world, DNA is ontologically and therefore ethically significant. Thus, transgenes from plants or animals considered impure or inappropriate as food may constitute a "contamination" of the foods into which they are transferred. This ontological understanding of DNA tends to be more strongly embraced by lay practitioners of these traditions than by the more scholarly theological and philosophical interpreters of the traditions.

4. Animal DNA is especially important: the focus groups interviewed in this project reflected an especially deep concern about the use of genetically modified animals for food as well as about the transfer of animal transgenes into other plants or species of animal. This should not be surprising since most, though not all, food sensitivities and prohibitions associated with the religious and moral traditions considered in this book are related to the raising, slaughtering, and consumption of animals as food. Many of these prohibitions and restrictions are grounded in considerations having to do with the ethical treatment of animals, as in the case of Buddhist, Hindu, and ethical vegetarian practices. Thus, developers of new food technologies and the regulators who approve

these for the market should be aware that biotechnologies involving the modification of food animals or the transfer of animal genes into plant foods are likely to be met with significant levels of concern, and probably rejection, by the sectors of the market represented by adherents to these and other religious and moral communities.

5. Adherents of these food traditions have a strong interest in access to information on the genetic profile of the plant and animal food products they purchase and consume. The information they require to exercise their values freely in the marketplace is not just the information whether or not a product contains genetically modified organisms; they also need access to specific information on the source of the transgenic material. The primary source of concern for these people is not that a food product contains GM material but rather the source or origin of that material.

The interest of these religious and moral communities should not be treated as one among many sets of preferences to be exercised in the marketplace. These interests rise to the level of fundamental rights of religious and moral conscience, to which liberal democratic society should ascribe special weight and respect.

Respect for these fundamental rights of religious and moral practice make it incumbent upon regulators of food biotechnology in these societies to establish mechanisms that require public access to the information about the origin of any transgenes in genetically modified products. Whether this is best accomplished by providing this information directly on the labels of all GM products or by simply providing to the consumer an apparent and easy means of obtaining this information, is a matter on which we make no judgment.

Notes

1. We owe this observation to David Waltner-Toews, made so powerfully in his delightful book *Food, Sex, and Salmonella: The Risk of Environmental Intimacy* (Toronto: NC, 1992).

2. See, for example, Martin Heidegger, "The Question concerning Technology," in *Basic Writings*, rev. ed. (New York: HarperCollins, 1993); Albert Borgmann, *Technology and the Character of Contemporary Life: A Philosophical Inquiry* (Chicago: University of Chicago Press, 1984).

3. Canadian regulations did not provide a basis for the denial of registration of BST on human health grounds, so the Canadian Food Inspection Agency, recognizing the strength of the public opposition, chose to ban it on the basis of impacts on animal health.

4. Industry promoters of biotechnology and many of the scientists involved in its development often claim that there is "nothing really new" in the technology of recombinant DNA, since the transfer of genes, even between species, happens naturally or has been accomplished for centuries.

5. See especially the chapter by Vasudha Narayanan as an example of the way in which the notion of purity shapes practices around food and eating in the Hindu religious culture.

6. An example is Bt corn, in which a gene is transferred from a naturally occurring soil bacterium (Bacillus thuringiensis) that produces a toxin lethal to the major pest in North American corn, the European corn borer. The bacterium itself is used as an organic pest control agent against the corn borer. When the requisite gene from the bacterium is spliced into Bt corn, the plant expresses the same toxin and hence becomes resistant to the corn borer. Bt corn needs no further chemical control against this pest.

7. As in the case of Round-Up Ready® corn, canola, soybeans, wheat, and other crops, which contain a transgene that confers resistance to a general herbicide (Round-Up) used for weed control.

8. See *Alliance for Bio-Integrity v. Shalala* 116 F. Supp. 2d 166 (2000), U.S. District Court for the District of Columbia.

9. According to these regulations, products that contain genetically modified organisms must be labeled as such. Products that contain less than 0.9 percent of content from GM ingredients are exempt from the labeling requirement, provided the content is adventitious. For a text of the EU regulation (Regulation (EC) 1830/2003, see http://europa.eu.int/eur-lex/pri/en/oj/dat/2003/l_268/l_26820031018en00010023.pdf.

10. Prevention of Food Adulteration Rules, Sections 37-E and 48-F.

11. For a summary of public opinion data, see e.g. R.W. Harrison and E. McLennon, "Analysis of U.S. Consumer Preferences for Labeling of Biotech Foods." Paper presented at 13th Annual World Food and Agribusiness Forum, 2003. http://www.ifama.org/conferences/2003Conference/papers/harrison.pdf.

12. C. A. Carter and G. P. Gruère, "Mandatory Labeling of Genetically Modified Foods: Does It Really Provide Consumer Choice?" *AgBioForum* 6, nos. 1&2 (2003), pp. 68–70. Available online at http://www.agbioforum.org.

13. See Section 2, Article 13, clause 2(b) of Regulation (EC) 1830/2003.

14. For further information on European Union regulations concerning GMO labeling, see the European Commission Food and Feed Safety website, http://europa.eu.int/comm/food/food/biotechnology/gmfood/labeling_en.htm.

15. See *Standards for Voluntary Labeling and Advertising of Foods That Are and Are Not Products of Genetic Engineering* (April 2004). http://www.pwgsc.gc.ca/cgsb/on_the_net/032_0315/032_0315_1995-e.pdf.

16. Ibid.

17. *Guidance for Industry: Voluntary Labeling Indicating Whether Foods Have or Have Not Been Developed Using Bioengineering. Draft Guidance.* U.S. Food and Drug Administration, Center for Food Safety and Nutrition, Washington DC. January 17, 2001. Docket Number 00D-1598. http://www.cfsan.fda.gov/~lrd/../~dms/biolabgu. html.

18. See, for example, Alaska's *Act Relating to Labeling and Identification of Genetically Modified Fish and Fish Products.* Senate Bill 25, online at http://www.legis. state.ak.us/basis/get_bill_text.asp?hsid=SB0025A&session=24.

19. E. Einsiedel, Consumers and GM Food Labels: Providing Information or Sowing Confusion? *AgBioForum* 3 (2000), pp. 231–35.

20. For discussion, see N. Kalaitzandonakes & J. Bijman, "Who Is Driving Biotechnology Acceptance?" *Nature Biotechnology* 21 (2003), pp. 366–69.

21. David Talbot, "Where's the Beef From?" *Technology Review* online at http:// beta.technologyreview.com/Biotech/13641.

22. http://usbiotechreg.nbii.gov/index.asp.

23. http://europa.eu.int/comm/food/dyna/gm_register/index_en.cfm.

24. R. Sherlock and Kawar, "Regulating Genetically Engineered Organisms: The Case of the Dairy Industry," in *Biotechnology: Assessing Social Impacts and Policy Implications,* D. J. Webber, ed. (New York: Greenwood, 1990), pp. 117–29; M. Nestle, "Food Biotechnology: Labeling Will Benefit Industry as Well as Consumers," *Nutrition Today* 13, no. 1 (1998), pp. 6–12.

25. The authors are grateful to Paul Thompson for his valuable assistance in drafting portions of the preceding paragraphs on the utilitarian and autonomy approaches to the labeling issue. Paul should not be held responsible for the overall argument, however.

26. The authors again acknowledge Paul Thompson for this point. See R. Chadwick, "Novel, Natural, Nutritious: Towards a Philosophy of Food," *Proceedings of the Aristotelian Society* (2000), 193–208; K. P. Rippe, "Novel Foods and Consumer Rights: Concerning Food Policy in a Liberal State," *Journal of Agricultural and Environmental Ethics* 12 (2000), pp. 71–80; H. Zwart, "A Short History of Food Ethics," *Journal of Agricultural and Environmental Ethics* 12 (2000), pp. 113–26; Robert Streiffer and Alan Rubel, "Democratic Principles and Mandatory Labeling of Genetically Engineered Food," *Public Affairs Quarterly* 18 (2004), p. 223.

Contributors

Samuel Abraham received his PhD in genetics from the University of British Columbia. He joined Inflazyme Pharmaceuticals as a division leader in cell and molecular biology. He also serves as a member of the scientific advisory board of AltaChem Pharma. Dr. Abraham is currently the director of the technology development office for the British Columbia Cancer Agency, where he manages the development of discoveries arising from research activities.

Donald Bruce has been director of the Society, Religion, and Technology Project (SRT) of the Church of Scotland since 1992. He has doctorates in chemistry and in theology, having formerly worked for fifteen years in nuclear energy in research, safety and risk regulation, and energy policy. He is coeditor with his wife, Ann, of *Engineering Genesis*, a study that has had a considerable impact in the developing debate on GM food in the UK and more widely. Dr. Bruce has been a member of the Scottish Science Advisory Committee and the public issues advisory group of the UK Biotechnology Research Council. He is an Honorary Fellow in Ethics at the Scottish Agricultural College and a member of the bioethics working groups of the Conference of European Churches. He is a frequent speaker, writer, and broadcaster on bioethical issues.

Conrad G. Brunk is professor of philosophy and past director of the Centre for Studies in Religion and Society at the University of Victoria. His areas of research and teaching include ethical and religious aspects of environmental and health risk perception and communication and value aspects of science in public policy. Dr. Brunk is a regular consultant to the Canadian government and international organizations on environmental and health risk management and biotechnology. He served as cochair of the Royal Society of Canada Expert Panel on the Future of Food Biotechnology and from

2002 through 2004 as a member of the Canadian Biotechnology Advisory Committee. He is coauthor with Lawrence Haworth and Brenda Lee of *Value Assumptions in Risk Assessment* (1991) and author of numerous articles in journals and books on ethical issues in technology, the environment, law, and professional practice. Professor Brunk holds a PhD in philosophy from Northwestern University.

Harold Coward is founding director and emeritus fellow of the Centre for Studies in Religion and Society, University of Victoria, Canada. He is a fellow of the Royal Society of Canada and currently serves as chair of the Ethics Advisory Committee of Genome British Columbia. A specialist in Indian philosophy and religion, he is author of eighteen books, including *Pluralism in the World Religions*, *The Philosophy of the Grammarians*, *Yoga and Psychology*, and most recently, *The Perfectibility of Human Nature in Eastern and Western Thought.* His current research is focused on the approach of religion to ethical issues in the genetic modification of plants and animals.

Hsiung Ping-chen is dean of the College of Liberal Arts at National Central University and a research fellow at the Institute of Modern History at Academia Sinica in Taiwan. Professor Hsiung is also an advisor to the Consortium of Humanities Centers and Institutes. A founder and coordinator of the Asian Humanities Network, Ping-chen promotes the study of late imperial China via interactive working groups using the now internationally recognized Ming-Ch'ing Studies Group website. Scholastically, Professor Hsiung pursues her research interests in child and family studies, traditional Chinese medicine, gender, and the construction of memory. Her most recent monograph, *A Tender Voyage: Children and Childhood in Late Imperial China*, reflects her devotion to advancing the field of Chinese studies. Her chapter in this volume represents the continuation of a line of her research combining China's culinary heritage, medical past, and genetically modified future.

Lyne Létourneau has been research professor at the Department of Animal Sciences at Laval University since 2002. She holds a doctorate in law from the University of Aberdeen (2000), as well as a master's in law (1993) and a bachelor's in law (1988) from the University of Montreal. Her research interests center on animal protection and agricultural biotechnology. In addition to articles related to animal law and ethics, she is author of *L'expérimentation animale: L'homme, l'éthique et la loi* (1994) and editor of *Bio-ingénierie et responsabilité sociale* (2006). She is coordinator of *Generistic* (www.generistic.org), a reference website on genetic engineering (2004).

David R. Loy is Best Professor of Religion and Ethics at Xavier University in Cincinnati, Ohio. His work is primarily in comparative philosophy and religion, particularly comparing Buddhist with modern Western thought. In addition to papers in various journals, he is the author of *Nonduality: A Study in Comparative Philosophy* (1988); *Lack and Transcendence: The Problem of Death and Life in Psychotherapy, Existentialism, and Buddhism* (1996); *A Buddhist History of the West: Studies in Lack* (2002); *The Great Awakening: A Buddhist Social Theory* (2003); and with Linda Goodhew, *The Dharma of Dragons and Daemons: Buddhist Themes in Modern Fantasy* (2004).

Lorenzo Mazgul is a PhD candidate in the Faculty of Land and Food Systems at the University of British Columbia. He lives in Victoria, BC.

Ebrahim Moosa is associate professor of Islamic studies in the Department of Religion at Duke University, and associate director (research) at the Duke Center for Islamic Studies. His interests span both classical and modern Islamic thought with a special interest in Islamic law, ethics, and theology. He is the author of *Ghazali and the Poetics of Imagination*, winner of the American Academy of Religion's Best First Book in the history of religions (2006), and editor of the last manuscript of the late Professor Fazlur Rahman, *Revival and Reform in Islam: A Study of Islamic Fundamentalism.* He was named Carnegie Scholar for 2005.

Vasudha Narayanan is distinguished professor in the Department of Religion at the University of Florida and a past president of the American Academy of Religion (2001-02). Her fields of interest are the Sri Vaishnava tradition; Hindu traditions in India, Cambodia, and America; Hinduism and the environment; and gender issues. She is currently working on Hindu temples and Vaishnava traditions in Cambodia. She is the author and editor of six books, including *Hinduism* (2004), and approximately one hundred articles, book chapters, and encyclopedia entries. Her new book, *A Hundred Autumns to Live: An Introduction to Hindu Traditions,* is forthcoming from Oxford University Press. Her research has been supported by grants and fellowships from several organizations, including the American Council of Learned Societies (2004-2005), National Endowment for the Humanities (1987, 1989-90, and 1998-99), the John Simon Guggenheim Foundation (1991-92), the American Institute of Indian Studies/ Smithsonian, and the Social Science Research Council.

Shiri Pasternak has a master's degree in cultural, social, and political thought from the University of Victoria. Her research was on genetically engineered food aid and the idea of "improvement" in colonial agrarian regimes since the seventeenth century. She was the associate director of the Forum on Privatization and the Public Domain in 2006. In this capacity, she organized a national interdisciplinary conference on "the commons," held in Victoria, British Columbia. She is now an independent researcher and writer living in Ottawa.

Nola M. Ries, MPA, LLM, is adjunct assistant professor in the School of Human and Social Development at the University of Victoria and a research associate with the Health Law Institute at the University of Alberta. Her work focuses primarily on legal and regulatory issues in health law, including regulation of new biotechnologies. She is involved in several Canadian research networks, including the Advanced Foods & Materials Network, where her research addresses legal, ethical, and social issues related to genomics, food, and health products. She is a lecturer in health law and regularly presents at Canadian and international legal, medical, and scientific conferences.

Leslie C. Rodgers is a senior consultant with the Praxis Group™, a western Canada-based firm specializing in public consultation, social science research, and processes to inform policy development. Leslie has managed processes for engaging the public in topical issues such as genomics, biotechnology, and health care; corporate responsibility toward the environment and community; and social issues such as youth unemployment and graffiti. Praxis is an innovator in process design through the use of such tools as Delphi panels, web surveys, study circles, and STS™ (web-based stakeholder tracking systems) to "translate theory into action." Ms. Rodgers resides in Vancouver, British Columbia.

Paul B. Thompson holds the W. K. Kellogg Chair in Agricultural, Food, and Community Ethics at Michigan State University, where he is also a professor in the Philosophy, Agricultural Economics and Community, Agriculture, and Recreation and Resource Studies departments. He has engaged in research and teaching on ethical issues associated with food production and consumption for twenty-five years and is the author or editor of seven books and more than one hundred journal articles and book chapters. He is a two-time recipient of the American Agricultural Economics Association Award for Excellence in Communication and serves as a member on numerous advisory committees, including Genome Canada's Science and Industry Advisory Committee.

Nancy Turner is a faculty member in the School of Environmental Studies at the University of Victoria, as well as a research associate with the Royal British Columbia Museum. She received her PhD from the Department of Botany at the University of British Columbia in 1974. She has authored and coauthored many papers and book chapters and more than twenty books in the general areas of ethnobotany, ethnoecology, traditional ecological knowledge, and sustainable resource use in Canada and British Columbia. Her most recent books are *Plants of Haida Gwaii* (2004), *The Earth's Blanket* (2005), *"Keeping It Living"* (2005; coedited with Douglas Deur), and *Plants of the Gitga'at People* (2006; coauthored with Judith Thompson). Nancy has particular interests in traditional food systems of indigenous peoples, in traditional land and resource management systems, in folk plant classification and botanical nomenclature, and in the acquisition and transmission of traditional ecological knowledge and practice in the context of a changing world.

Laurie Zoloth is director of the Center for Bioethics, Science, and Society and professor of medical ethics and humanities at Feinberg School of Medicine and professor of religion in the Jewish studies faculty at Weinberg College of Arts and Science, Northwestern University. She directs bioethics at the Center for Genetic Medicine, the Center for Regenerative Medicine, and the Institute for Nanotechnology. She has published extensively in the areas of ethics, family, feminist theory, religion and science, Jewish studies, and social policy. Her current research projects include work on emerging issues in medical and research genetics, ethical issues in stem cell research, and distributive justice in health care.

Index